CBEST Math 10 Practice Tests

Realistic Full-Length Test and Detailed Explanations to Questions.

Ultimate Companion to Textbooks and Workbooks for Ultimate

CBEST Math Prep and Study Review.

Dr. Abolfazl Nazari

Copyright © 2024 Dr. Abolfazl Nazari

PUBLISHED BY EFFORTLESS MATH EDUCATION

EFFORTLESSMATH.COM

All rights reserved. No part of this publication may be reproduced, distributed, or transmitted in any form or by any means, including photocopying, recording, or other electronic or mechanical methods, without the prior written permission of the author, except in the case of brief quotations embodied in critical reviews and certain other noncommercial uses permitted by copyright law, including Section 107 or 108 of the 1976 United States Copyright Act.

Copyright ©2024

Welcome to CBEST Math 10 Practice Tests

2024

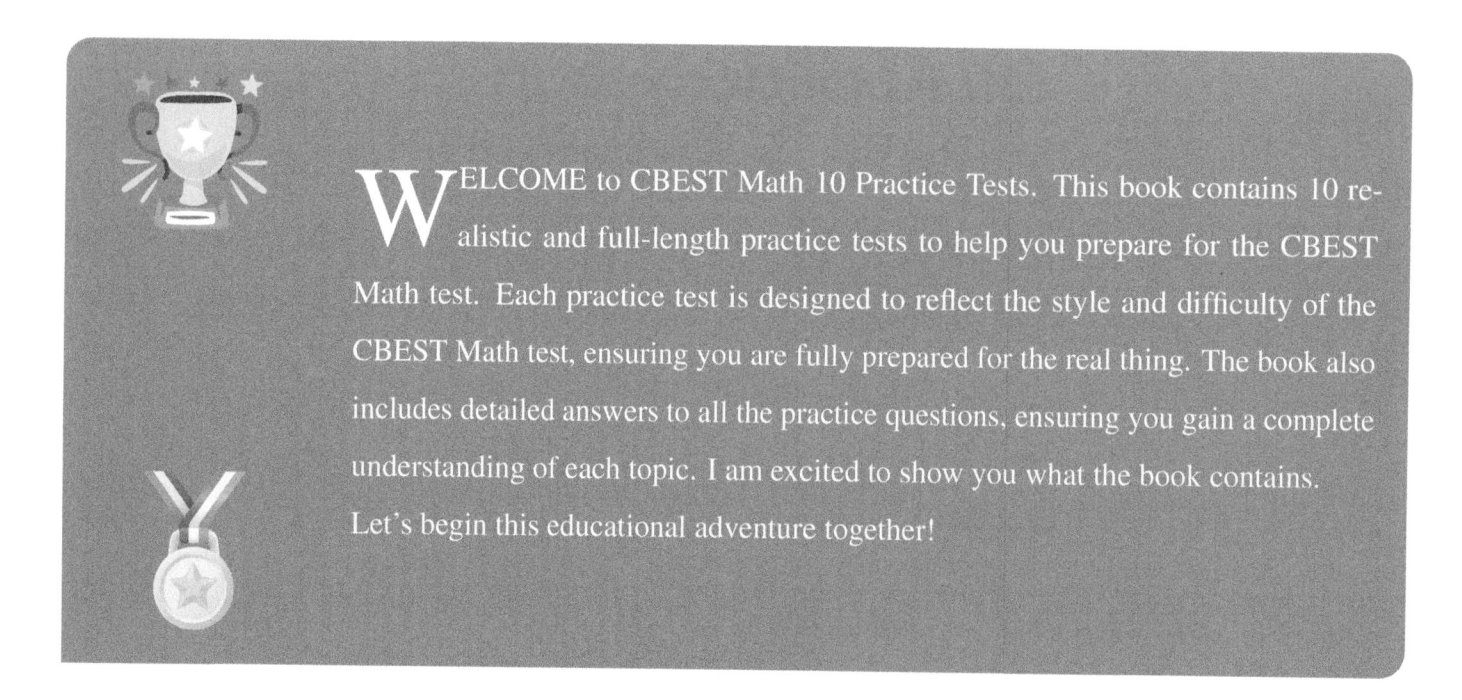

WELCOME to CBEST Math 10 Practice Tests. This book contains 10 realistic and full-length practice tests to help you prepare for the CBEST Math test. Each practice test is designed to reflect the style and difficulty of the CBEST Math test, ensuring you are fully prepared for the real thing. The book also includes detailed answers to all the practice questions, ensuring you gain a complete understanding of each topic. I am excited to show you what the book contains. Let's begin this educational adventure together!

CBEST Math 10 Practice Tests is structured into 10 chapters, each containing a realistic and full-length *Practice Test*. Each chapter ends with detailed answers to all the practice questions, ensuring you gain a complete understanding of each topic. The content is tailored to prepare you for the CBEST Math test.

The book also includes CBEST Math test tips and strategies. These are designed to help you understand the best way to approach each question type and pass the test. The book is designed to be used as a standalone practice tests, but it can also be used as a supplement to a CBEST Math course or as a companion to a CBEST Math study guide.

What is included in this book

- ☑ Online resources for additional practice and support.
- ☑ CBEST Math test tips and strategies.
- ☑ A guide on how to use this book effectively.
- ☑ 10 realistic and full-length practice tests with detailed answers.

Effortless Math's CBEST Online Center

Effortless Math Online CBEST Center offers a complete study program, including the following:

- ☑ *Step-by-step instructions on how to prepare for the CBEST Math test*
- ☑ *Numerous CBEST Math worksheets to help you measure your math skills*
- ☑ *Complete list of CBEST Math formulas*
- ☑ *Video lessons for all CBEST Math topics*
- ☑ *Full-length CBEST Math practice tests*

Visit EffortlessMath.com/CBEST to find your online CBEST Math resources.

Scan this QR code

(No Registration Required)

How to Prepare for the AFOQT Math Test

Preparing for the AFOQT Math test may initially seem daunting and unexciting. However, there are proven methods to make your study journey more effective and engaging. In this section, we present a structured six-step program designed to streamline your AFOQT Math preparation process, ensuring efficiency and reducing the feeling of overwhelm.

▷ **Step 1** - *Develop a comprehensive study plan*

▷ **Step 2** - *Curate your study materials carefully*

▷ **Step 3** - *Engage in review, learning, and practice*

▷ **Step 4** - *Master effective test-taking strategies*

▷ **Step 5** - *Familiarize yourself with the AFOQT Math test structure and take practice tests*

▷ **Step 6** - *Evaluate and reflect on your performance*

Step 1: Develop a comprehensive study plan

Having a plan makes accomplishing tasks much easier. When it comes to preparing for the AFOQT Math test, a study plan can help you stay organized. Take some time to create a study plan that fits into your daily life, including your schoolwork and other responsibilities. Allocate enough time each day for studying, and consider breaking the test into smaller sections, focusing on one topic at a time. Preparing for the AFOQT Math test is a personalized journey, and your study plan should be tailored to your individual needs and learning style.

Here are some practical steps to help you create a study plan that suits you:

- **Understand Your Learning Style and Study Habits:** We all have unique ways of learning. Embrace your individuality and think about what works best for you. Do you prefer using AFOQT Math prep books, or do a mix of textbooks and video lessons suit you better? Are you more productive studying for short sessions every night, or is a morning study routine more effective for you before heading to work or school?

- **Assess Your Available Time:** Take a close look at your current schedule to determine how much time you can consistently dedicate to AFOQT Math study.

- **Craft Your Study Schedule:** Now, incorporate your study schedule into your daily calendar, treating it as seriously as any other commitment. Schedule specific study, practice, and review sessions. Plan which topics you'll tackle on specific days to ensure you cover each concept adequately. Develop a thoughtful, realistic, and adaptable study plan.

- **Stick to Your Schedule:** A study plan becomes effective when you follow it consistently. Strive to develop a schedule that you can maintain throughout your study program.

- **Regularly Review and Adjust Your Study Plan:** Life may introduce new commitments, so it's important to be flexible. Periodically check in with yourself to ensure you're staying on track with your study plan. Remember, the primary goal is to adhere to your plan. If you find that it's not yielding the results you desire, don't be disheartened. Feel free to make adjustments as you discover what works best for you.

Step 2: Curate your study materials carefully

The abundance of textbooks and online materials available for AFOQT Math preparation might seem overwhelming at first. But fear not! We got you covered with everything you need to excel in your AFOQT Math test. We provide various resources, including study guides, workbooks, and practice tests, all designed to help you succeed.

In addition to the book's content, you'll have access to Effortless Math's online resources, including video lessons, worksheets, formulas, and more. For even more online AFOQT Math resources, you can visit **EffortlessMath.com/AFOQT**. With these resources at your fingertips, you'll have all the tools you need to succeed in your AFOQT Math journey.

Step 3: Engage in review, learning, and practice

To effectively prepare for the AFOQT Math test, start by solidifying your understanding of essential math concepts, then expand your knowledge by addressing weaker areas with resources like textbooks and online tutorials. This approach ensures a solid foundation and a comprehensive grasp of the material.

Incorporate extensive practice into your study routine, tackling a wide range of problem types to refine your problem-solving skills and get accustomed to the test format. Simulate exam conditions to improve time management and lessen anxiety. Regular review of your practice sessions to identify and work on weaknesses will enhance your confidence and ability, setting you up for success on the test.

Step 4: Master effective test-taking strategies

Within the upcoming sections, you'll discover invaluable techniques and advice designed to enhance your test-taking skills and potentially increase your score. You'll gain insights on strategic thinking and when it's advantageous to make educated guesses if a question perplexes you. Implementing these AFOQT Math *test-taking strategies* and tips can elevate your performance and lead to success on the exam. Be sure to apply these strategies when tackling *practice tests* to build your confidence.

Step 5: Familiarize yourself with the AFOQT Math test structure and take *practice tests*

You need to know your AFOQT Math test inside and out to perform at your best. This book provides a comprehensive overview of the AFOQT Math test, including its structure, question types, and time limits. Make sure to read through this section carefully to familiarize yourself with the test's format and requirements.

Once you've absorbed the instructions and lessons and feel adequately prepared, make the most of the two full-length AFOQT Math included in this book. These practice tests closely mirror the actual AFOQT Math test format. To get the most out of them, create a testing environment similar to the real exam. Find a quiet place, set a timer, and work through as many questions as you can within the allotted time. After completing each *practice test*, delve into the detailed answer explanations provided. These explanations will help you pinpoint your weaker areas, learn from your errors, and ultimately improve your AFOQT Math score.

Step 6: Evaluate and reflect on your performance

Once you've completed the *practice tests*, carefully review the answer keys and explanations. This process will help you identify the questions you answered correctly and those you may have missed. Remember, making a few mistakes is a natural part of learning, so don't feel disheartened. Instead, view these errors as valuable opportunities for improvement. They serve as indicators of your strengths and areas that may need more attention.

Your test results can serve as a valuable tool to assess your readiness for the actual AFOQT Math test. You can use them to gauge whether you require additional practice or if you've reached the point where you feel confident in tackling the real exam.

Tips for Making the Most of This Book

This book includes 10 chapters, each containing a realistic, full-length practice test. These practice tests are designed to help you prepare for your final exam. The tests are followed by answer keys, and at the end of each chapter, you can find detailed solutions to all the practice questions.

To get the most out of these practice tests, remember to time yourself. This will help you get used to the time pressure of a real exam. In your actual CBEST Math test, time management is crucial. You will need to answer questions quickly and accurately.

After completing the practice tests, check your performance. We have included an answer key for each test. Please mark your answers and compare them with the answer key. If an answer does not match, be sure to review the explanation provided. This will help you understand how to solve similar problems in the future.

If you need further assistance, you can use the study guides in this series. We provide various study guides, including *CBEST Math in 30 Days*, *CBEST Math in 10 Days*, and *CBEST Math Made Easy*. These study guides are designed to help you grasp the concepts and strategies needed to pass the CBEST Math test.

In addition to the material covered in this book and in the series, it is essential to have a solid plan for your test preparation. Consider the following tips:

- **Begin Exam Preparation Early.** Avoid last-minute cramming by starting your study sessions well in advance, preferably at least a week before the exam. This allows ample time for thorough review and practice.
- **Consistent Daily Study Sessions.** Opt for regular study periods of 30 to 45 minutes each day instead of prolonged, last-minute studying. This consistent approach aids in better retention and reduces stress.
- **Engage in Active Note-Taking.** Jotting down notes is an effective way to internalize key concepts and maintain focus. Regularly review your notes to reinforce your understanding.
- **Revisit Challenging Topics.** Allocate extra time to review topics you find difficult. Revisiting these topics frequently will help solidify your understanding and improve your performance.
- **Emphasize Practice.** Ensure you engage in ample practice. The end-of-chapter problems and additional practice workbooks in the series are valuable resources for exam preparation.

Contents

1	**CBEST Test Review and Strategies**	1
1.1	The CBEST Test Review	1
1.2	CBEST Math Test-Taking Strategies	2
2	**Practice Test 1**	6
2.1	Practices	6
2.2	Answer Keys	23
2.3	Answers with Explanation	24
3	**Practice Test 2**	36
3.1	Practices	36
3.2	Answer Keys	58
3.3	Answers with Explanation	59
4	**Practice Test 3**	68
4.1	Practices	68
4.2	Answer Keys	87
4.3	Answers with Explanation	88

5	Practice Test 4	95
5.1	Practices	95
5.2	Answer Keys	113
5.3	Answers with Explanation	114

6	Practice Test 5	122
6.1	Practices	122
6.2	Answer Keys	141
6.3	Answers with Explanation	142

7	Practice Test 6	151
7.1	Practices	151
7.2	Answer Keys	165
7.3	Answers with Explanation	166

8	Practice Test 7	176
8.1	Practices	176
8.2	Answer Keys	189
8.3	Answers with Explanation	190

9	Practice Test 8	203
9.1	Practices	203
9.2	Answer Keys	217
9.3	Answers with Explanation	218

10	Practice Test 9	232
10.1	Practices	232
10.2	Answer Keys	246

10.3	Answers with Explanation	247

11 Practice Test 10 259

11.1	Practices	259
11.2	Answer Keys	272
11.3	Answers with Explanation	273

1. CBEST Test Review and Strategies

1.1 The CBEST Test Review

The *California Basic Educational Skills Test (CBEST)* is a standardized test administered to individuals who aspire to become educators within California. The CBEST is crafted to evaluate fundamental reading, writing, and mathematical skills deemed essential for the profession. Special emphasis on the *CBEST Mathematics section* is paramount, as it reflects the educator's ability to solve problems, interpret data, and convey mathematical concepts effectively.

The Mathematics portion of the CBEST encompasses three primary domains:

- **Estimation, Measurement, and Statistical Principles**
- **Computation and Problem Solving**
- **Numerical and Graphic Relationships**

Comprising 50 multiple-choice questions, the CBEST Math test assesses the candidate's aptitude in applying arithmetic operations, understanding principles of estimation and measurement, and utilizing statistical information. Calculators are not permitted during this test, emphasizing the importance of a strong mental arithmetic capability and a thorough understanding of elementary mathematical concepts.

Scoring for the CBEST involves achieving a minimum score of 41 in each section, with a passing total score of 123 across all three sections combined. The precision of the assessment is fine-tuned with a scoring system that reflects a candidate's performance accurately.

Given the no-retake rule of the CBEST once passed, and the crucial role it plays in obtaining teaching credentials, exhaustive preparation is indispensable. This guide includes detailed study materials and practice tests, specifically for the *Mathematics section*, to help candidates prepare effectively. These practice tests are carefully designed to emulate the structure and content of the actual CBEST Math test, allowing for a realistic and focused review.

1.2 CBEST Math Test-Taking Strategies

Successfully navigating the CBEST Math test requires not only a solid understanding of mathematical concepts but also effective problem-solving strategies. In this section, we explore a range of strategies to optimize your performance and outcomes on the CBEST Math test. From comprehending the question and using informed guessing to finding ballpark answers and employing backsolving and numeric substitution, these strategies will empower you to tackle various types of math problems with confidence and efficiency.

#1 Understand the Questions and Review Answers

Below are a set of effective strategies to optimize your performance and outcomes on the CBEST Math test.

- **Comprehend the Question:** Begin by carefully reviewing the question to identify keywords and essential information.
- **Mathematical Translation:** Translate the identified keywords into mathematical operations that will enable you to solve the problem effectively.
- **Analyze Answer Choices:** Examine the answer choices provided and identify any distinctions or patterns among them.
- **Visual Aids:** If necessary, consider drawing diagrams or labeling figures to aid in problem-solving.
- **Pattern Recognition:** Look for recurring patterns or relationships within the problem that can guide your solution.
- **Select the Right Method:** Determine the most suitable strategies for answering the question, whether it involves straightforward mathematical calculations, numerical substitution (plugging in numbers), or testing the answer choices (backsolving); see below for a comprehensive explanation of these methods.
- **Verification:** Before finalizing your answer, double-check your work to ensure accuracy and completeness.

1.2 CBEST Math Test-Taking Strategies

Let's review some of the important strategies in detail.

#2 Use Educated Guessing

This strategy is particularly useful for tackling problems that you have some understanding of but cannot solve through straightforward mathematics. In such situations, aim to eliminate as many answer choices as possible before making a selection. When faced with a problem that seems entirely unfamiliar, there's no need to spend excessive time attempting to eliminate answer choices. Instead, opt for a random choice before proceeding to the next question.

As you can see, employing direct solutions is the most effective approach. Carefully read the question, apply the math concepts you've learned, and align your answer with one of the available choices. Feeling stuck? Make your best-educated guess and move forward.

Never leave questions unanswered! Even if a problem appears insurmountable, make an effort to provide a response. If necessary, make an educated guess. Remember, you won't lose points for an incorrect answer, but you may earn points for a correct one!

#3 Ballpark Estimates

A *"ballpark estimate"* is a *rough approximation*. When dealing with complex calculations and numbers, it's easy to make errors. Sometimes, a small decimal shift can turn a correct answer into an incorrect one, no matter how many steps you've taken to arrive at it. This is where ballparking can be incredibly useful.

If you have an idea of what the correct answer might be, even if it's just a rough estimate, you can often eliminate a few answer choices. While answer choices typically account for common student errors and closely related values, you can still rule out choices that are significantly off the mark. When facing a multiple-choice question, deliberately look for answers that don't even come close to the ballpark. This strategy effectively helps eliminate incorrect choices during problem-solving.

#4 Backsolving

A significant portion of questions on the CBEST Math test are presented in multiple-choice format. Many test-takers find multiple-choice questions preferable since the correct answer is among the choices provided.

Typically, you'll have four options to choose from, and your task is to determine the correct one. One effective approach for this is known as *"backsolving."*

As mentioned previously, solving questions directly is the most optimal method. Begin by thoroughly examining the problem, calculating a solution, and then matching the answer with one of the available choices. However, if you find yourself unable to calculate a solution, the next best approach involves employing *"backsolving."*

When employing backsolving, compare one of the answer choices to the problem at hand and determine which choice aligns most closely. Frequently, answer choices are arranged in either ascending or descending order. In such cases, consider testing options B or C first. If neither is correct, you can proceed either up or down from there.

#5 Plugging In Numbers

Using numeric substitution or *'plugging in numbers'* is a valuable strategy applicable to a wide array of math problems encountered on the CBEST Math test. This approach is particularly helpful in simplifying complex questions, making them more manageable and comprehensible. By employing this strategy thoughtfully, you can arrive at the solution with ease.

The concept is relatively straightforward. Simply replace unknown variables in a problem with specific values. When selecting a number for substitution, consider the following guidelines:

- Opt for a basic number (though not overly basic). It's generally advisable to avoid choosing 1 (or even 0). A reasonable choice often includes selecting the number 2.
- Avoid picking a number already present in the problem statement.
- Ensure that the chosen numbers are distinct when substituting at least two of them.
- Frequently, the use of numeric substitution helps you eliminate some of the answer choices, so it's essential not to hastily select the first option that appears to be correct.
- When faced with multiple seemingly correct answers, you may need to opt for a different set of values and reevaluate the choices that haven't been ruled out yet.
- If your problem includes fractions, a valid solution might require consideration of either *the least common denominator (LCD)* or a multiple of the LCD.
- When tackling problems related to percentages, it's advisable to select the number 100 for numeric substitution.

It is Time to Test Yourself

It's time to refine your skills with a practice examination designed to simulate the CBEST Math Test. Engaging with the practice tests will help you to familiarize yourself with the test format and timing, allowing for a more effective test day experience. After completing a test, use the provided answer key to score your work and identify areas for improvement.

Before You Start

To make the most of your practice test experience, please ensure you have:

- A pencil for marking answers on the answer sheet.
- A timer to manage pacing, replicating potential time constraints in other testing scenarios.

Please note the following important points as you prepare to take your practice test:

- It's okay to guess! There is no penalty for incorrect answers, so make sure to answer every question.
- After completing the test, review the answer key to understand any mistakes. This review is crucial for your learning and preparation.
- An answer sheet is provided for you to record your answers. Make sure to use it.
- For each multiple-choice question, you will be presented with possible choices. Your task is to choose the best one.

Good Luck! Your preparation and practice are the keys to success.

2. Practice Test 1

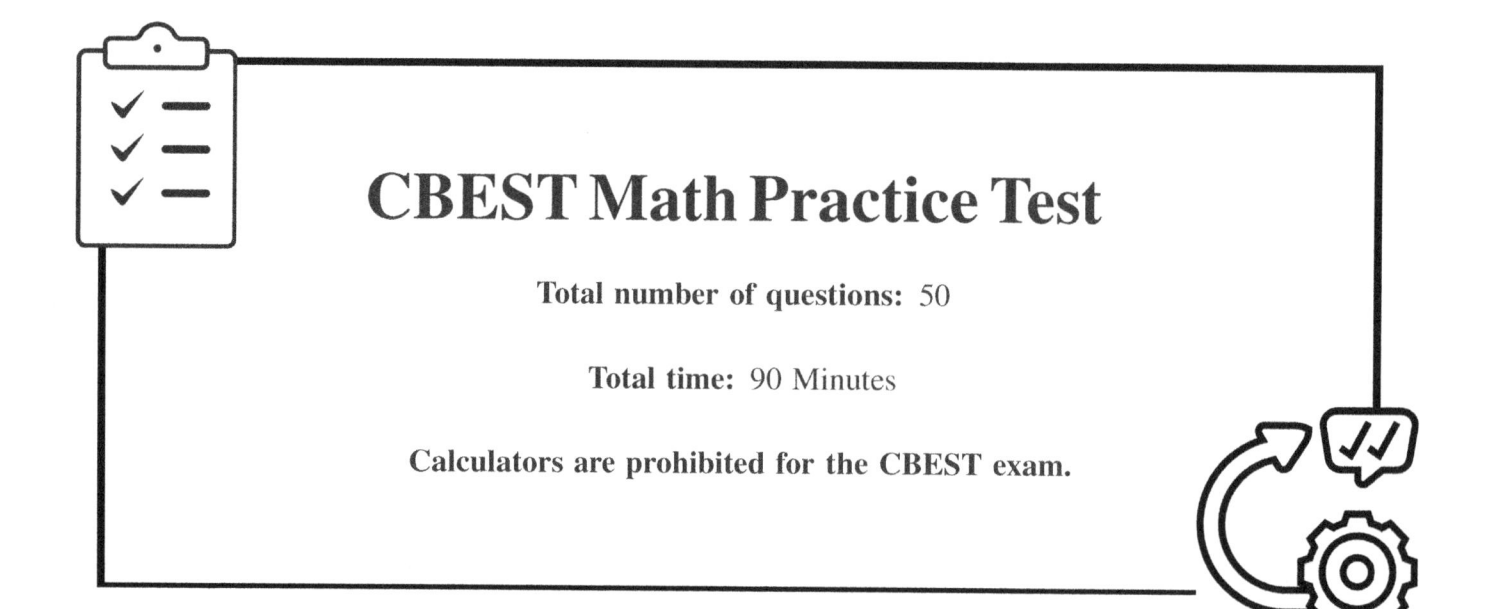

CBEST Math Practice Test

Total number of questions: 50

Total time: 90 Minutes

Calculators are prohibited for the CBEST exam.

2.1 Practices

1) A cone has a height of 15 *cm* and a radius of 4 *cm*. What is the volume of the cone in cubic centimeters? (Use $\pi = 3.14$).

2.1 Practices

☐ A. 240.6 cm^3

☐ B. 251.2 cm^3

☐ C. 264.4 cm^3

☐ D. 380.5 cm^3

☐ E. 396.3 cm^3

2) Lila is considering investing $3,000 in a new savings account. She is comparing two different options:

• Account X offers a 2.8% simple annual interest rate.

• Account Y offers a 3% interest rate compounded annually.

Assuming she won't make any additional deposits or withdrawals for 5 years, which account will yield more interest and by how much? (Round to nearest integer)

☐ A. Account X would earn Lila about $50 more interest than Account Y.

☐ B. Account Y would earn Lila about $50 more interest than Account X.

☐ C. Account X would earn Lila about $84 more interest than Account Y.

☐ D. Account Y would earn Lila about $58 more interest than Account X.

☐ E. Account Y would earn Lila about $150 more interest than Account X.

3) If the electricity cost for operating a machine is directly proportional to the number of hours it runs, the cost for running it for 120 hours was $60. What would be the cost for running it for 60 hours?

☐ A. $15

☐ B. $30

☐ C. $45

☐ D. $60

☐ E. $90

4) A linear function is determined by the points $(5,3)$ and $(-3,-5)$. What is the y-intercept of this linear function?

- A. −1.2
- B. −1.5
- C. −1.8
- D. −2
- E. −3

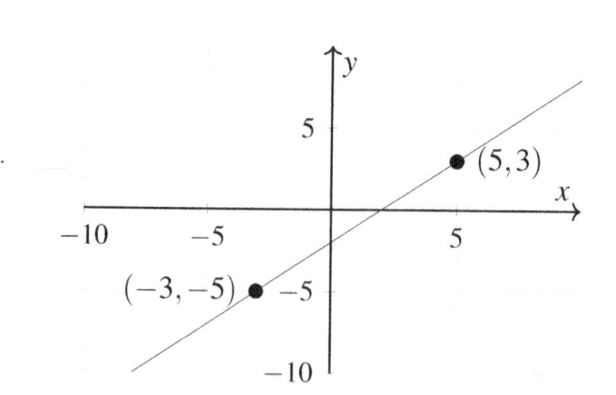

5) A rectangle is graphed on a coordinate grid with vertices at $(3,2)$, $(3,6)$, $(8,6)$, and $(8,2)$. The rectangle will be translated p units to the right and q units down. Which is the coordinates of the top-right vertex of the new rectangle after this translation?

- A. $(8+p, 6-q)$
- B. $(8+p, 2-q)$
- C. $(3+p, 2-q)$
- D. $(3+p, 6-q)$
- E. $(8+p, 2-q)$

6) Susan's job is to deliver packages for a company. Each month she receives a fixed salary plus extra money for each package delivered.
 - In August, Susan delivered 500 packages and received a total of $3,500.
 - In September, she delivered 800 packages and received a total of $4,400.

What function can be used to find y, the total amount she earns in a month if she delivers x packages?

- A. $y = 4.00x$
- B. $y = 3.00x$
- C. $y = 3.00x + 2,000$
- D. $y = 2,000x + 3.00$
- E. $y = 2,000x$

7) A craftsman creates a cylindrical pottery piece with a height of 12 inches and a diameter of 10 inches. Which equation can be used to find V, the volume of the pottery piece in cubic inches?

- A. $V = \pi(10)^2 \frac{(12)}{2}$
- B. $V = \pi(5)^2(12)$

2.1 Practices

- [ ] C. $V = \pi(5)\left(\frac{12}{2}\right)$
- [ ] D. $V = \pi(10)^2(12)$
- [ ] E. $V = \pi(5)(12)$

8) Which measurements could represent the side lengths in meters of a right triangle?

- [ ] A. 2 m, 5 m, 6 m
- [ ] B. 10 m, 10 m, 10 m
- [ ] C. 7 m, 9 m, 11 m
- [ ] D. 3 m, 4 m, 8 m
- [ ] E. 8 m, 15 m, 17 m

9) The graph of a quadratic function is shown on the grid. Which equation best represents the axis of symmetry?

- [ ] A. $y = -2x + 4$
- [ ] B. $x = 2$
- [ ] C. $x = 3$
- [ ] D. $y = 2$
- [ ] E. $x = -2$

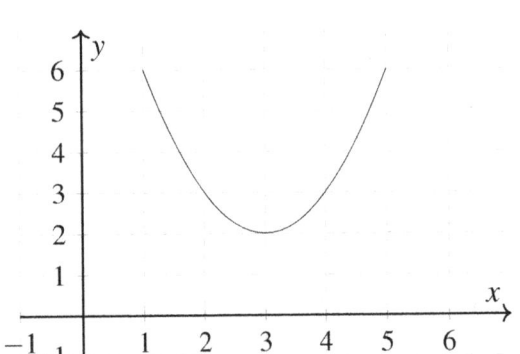

10) Which graph best represents the solution set of $x > 2y - 5$?

- [ ] A. 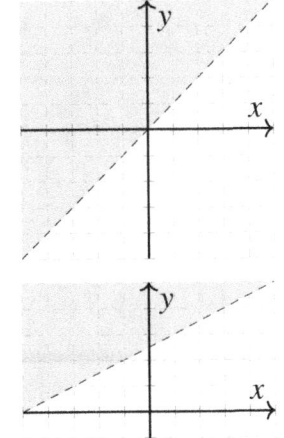
- [ ] B.
- [ ] C.
- [ ] D.

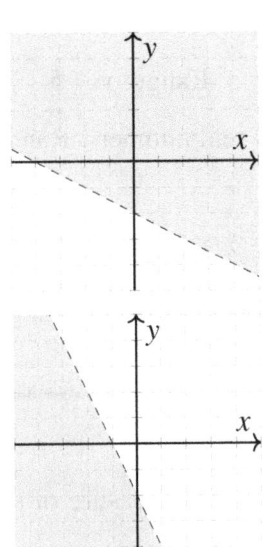

☐ E.

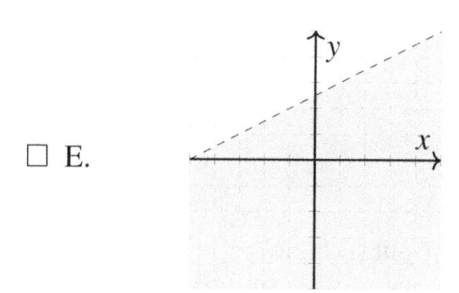

11) A local bookstore tracks the relationship between the number of books in stock and the number of daily sales. The data is summarized in a table, suggesting a linear relationship. If the bookstore increases its inventory to 5,000 books, what is the best prediction for the number of daily sales based on the linear trend?

☐ A. 120

☐ B. 135

☐ C. 150

☐ D. 165

☐ E. 180

Books in Stock	Daily Sales
1000	60
2000	75
3000	90
4000	105

12) A quadratic function $f(x)$ is graphed on the coordinate plane. The graph of the function opens downwards and its vertex is at the point $(3,5)$. Which answer choice best represents the domain and range of this function?

☐ A. Domain: All real numbers; Range: $y \leq 5$

☐ B. Domain: $x \geq 3$; Range: $y \leq 5$

☐ C. Domain: All real numbers; Range: $y \geq 5$

☐ D. Domain: $x \leq 3$; Range: $y \geq 5$

☐ E. Domain: All real numbers; Range: $y = 5$

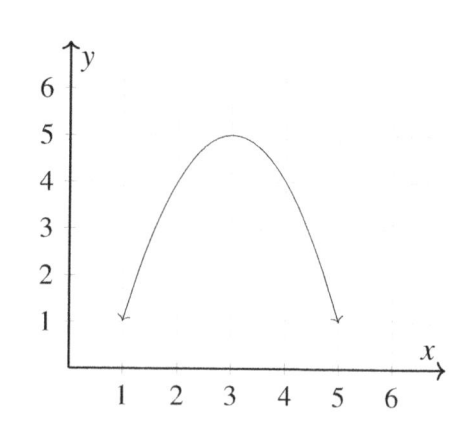

13) In a parallelogram, the measure of one angle is given as $(2x+10)°$ and the adjacent angle is given as $(3x-5)°$. Find the value of x.

2.1 Practices

☐ A. 55

☐ B. 50

☐ C. 45

☐ D. 40

☐ E. 35

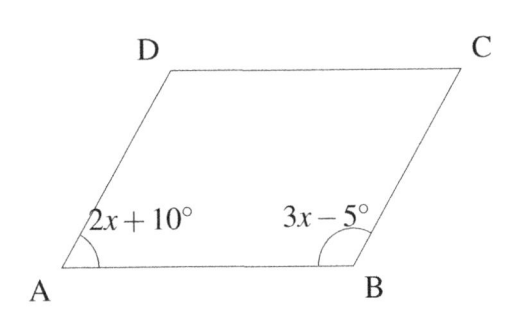

14) The graph of a line passes through the points $(1,4)$ and $(4,1)$. What is the rate of change of y with respect to x for this line?

☐ A. -2

☐ B. -1

☐ C. -3

☐ D. 2

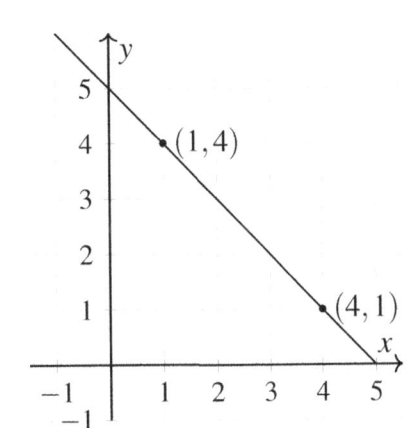

☐ E. 1

15) Consider two lines L_1 and L_2 with the following points: Line L_1 passes through $(2,-5)$, $(4,-4)$, and $(6,-3)$, Line L_2 passes through $(-4,7)$, $(0,3)$, and $(2,1)$. Which system of equations is represented by lines L_1 and L_2?

☐ A. $\begin{cases} x - 2y = 12, \\ x + y = 3 \end{cases}$

☐ B. $\begin{cases} 2x - y = 6, \\ 2x + y = 3 \end{cases}$

☐ C. $\begin{cases} 3x - 2y = 12, \\ 2x + 3y = 6 \end{cases}$

☐ D. $\begin{cases} x - 4y = 16, \\ 2x + y = 6 \end{cases}$

☐ E. $\begin{cases} 4x - 2y = 5, \\ 2x - y = 3 \end{cases}$

16) A quadratic function's graph is displayed in the coordinate plane. Which of the following equations best describes the graph?

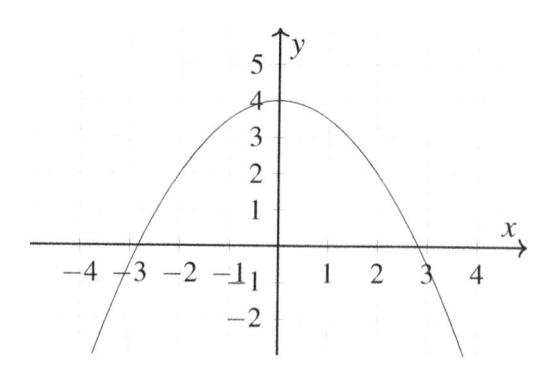

- A. $y = -2x^2 + 4$
- B. $y = -\frac{1}{2}x^2 + 4$
- C. $y = -\frac{1}{2}x^2 - 4$
- D. $y = \frac{1}{2}x^2 - 4$
- E. $y = 2x^2 - 4$

17) A hill in the countryside can be modeled as a cone with a diameter of 12 meters and a height of 6 meters. What is the closest measurement to the volume of the hill in cubic meters?

- A. 91 m^3
- B. 152 m^3
- C. 226 m^3
- D. 254 m^3
- E. 406 m^3

18) Which set of ordered pairs represents a function?

- A. $\{(3,6), (2,5), (1,3), (3,4)\}$
- B. $\{(0,1), (0,2), (1,3), (2,4)\}$
- C. $\{(7,3), (8,4), (9,5), (7,6)\}$
- D. $\{(4,-1), (5,-2), (6,-3), (7,-4)\}$
- E. $\{(-2,9), (-3,8), (-4,7), (-2,6)\}$

19) The graph of a linear function is shown on a grid. What is the zero of the function?

- A. -1
- B. 1
- C. $-\frac{1}{2}$
- D. $-\frac{1}{3}$
- E. $-\frac{1}{4}$

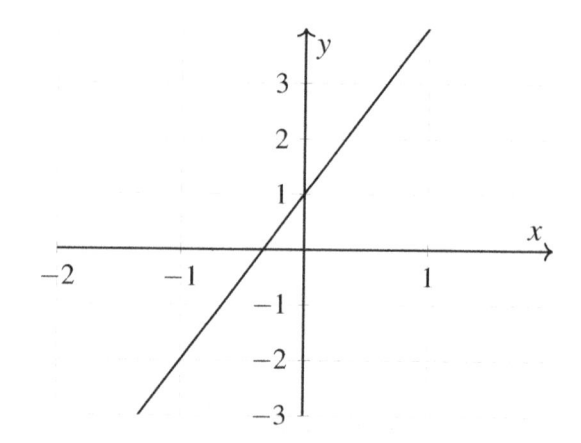

20) Among the following options, which one represents a non-proportional linear relationship between x and

2.1 Practices

y?

- ☐ A. A line passing through the origin and having a positive slope.
- ☐ B. A horizontal line intersecting the y-axis at y = 5.
- ☐ C. A vertical line intersecting the x-axis at x = −3.
- ☐ D. A line with a negative slope that does not pass through the origin.
- ☐ E. A line with a positive slope that passes through the points (2, 4) and (4, 8).

21) A customer needs to borrow $10,000 for a home renovation. Which loan option would result in the least amount of total interest paid over the loan period?

- ☐ A. A 2-year loan with a 9.5% annual simple interest rate.
- ☐ B. A 3-year loan with a 7.5% annual simple interest rate.
- ☐ C. A 4-year loan with a 5.8% annual simple interest rate.
- ☐ D. A 5-year loan with a 5.0% annual simple interest rate.
- ☐ E. A 6-year loan with a 4.8% annual simple interest rate.

22) The graph models the value of a computer over a 10-year period. Which equation best represents the relationship between x, the age of the computer in years, and y, the value of the computer in dollars over this 10-year period?

- ☐ A. $y = 150x + 9,000$
- ☐ B. $y = -0.003x + 9,000$
- ☐ C. $y = 200x + 7,000$
- ☐ D. $y = -0.002x + 9,000$
- ☐ E. $y = -200x + 9,000$

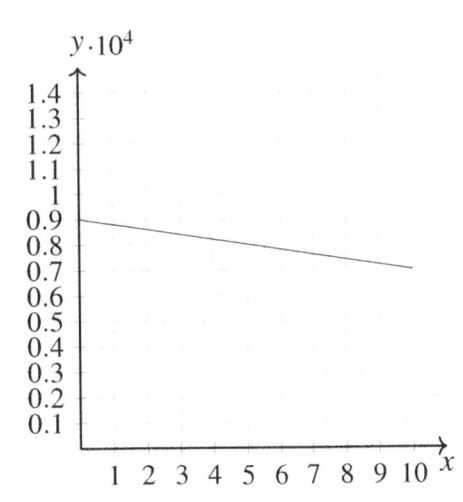

23) A globe is shaped like a sphere with a diameter of 20 centimeters. Which measurement is closest to the volume of the globe in cubic centimeters? (Use $\pi = 3.14$).

- ☐ A. $1,047.20\ cm^3$
- ☐ B. $2,094.40\ cm^3$
- ☐ C. $3,351.03\ cm^3$

☐ D. 4,186.67 cm^3

☐ E. 5,026.55 cm^3

24) A cargo container shaped like a rectangular prism has a base that measures 2 meters by 5 meters. The total surface area of the container is 90 square meters. What is the height of the container in meters?

☐ A. 2 m

☐ B. 3 m

☐ C. 4 m

☐ D. 5 m

☐ E. 6 m

25) A cyclist covers a distance of 104 kilometers in 208 minutes. Which function has a slope that best represents this rate of travel?

☐ A. $y = 0.5x$

☐ B. $y = 0.4x$

☐ C. $y = x$

☐ D. $y = 2x$

☐ E. $y = 3x$

26) The graph of a linear function is shown on the grid, passing through points $(2,4)$ and $(-2,-2)$. What are the slope and the y-intercept of this line?

☐ A. Slope $= -1.5$, y-intercept $= 1$

☐ B. Slope $= -1$, y-intercept $= 4$

☐ C. Slope $= 1$, y-intercept $= -2$

☐ D. Slope $= 1.5$, y-intercept $= 1$

☐ E. Slope $= 2$, y-intercept $= 0$

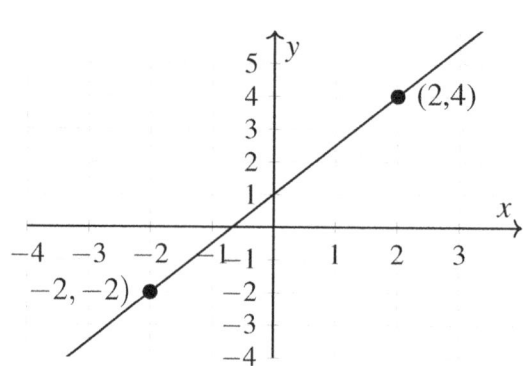

27) A garden's water tank initially contains 80 liters of water. The water level in the tank increases at a steady rate of 2 liters per day. Which function can be used to find T, the total number of liters of water in the tank after D days?

☐ A. $T = -80D + 2$

2.1 Practices

☐ B. $T = -2D + 80$

☐ C. $T = 2D - 80$

☐ D. $T = 2D + 80$

☐ E. $T = 80D + 2$

28) The table below shows the diagonal lengths of various screens in a store, measured in inches.

Screen	Length (inches)
A	$\frac{15}{4}$
B	$\frac{10}{3}$
C	$\sqrt{5}$
D	$\sqrt{20}$
E	$\frac{\sqrt{16}}{4}$

Which screen has the shortest diagonal length in inches?

☐ A. Screen A

☐ B. Screen B

☐ C. Screen C

☐ D. Screen D

☐ E. Screen E

29) The population of an island is 4,580,000 people. How is this number of people written in scientific notation?

☐ A. 4.58×10^{-6}

☐ B. 4.58×10^{-3}

☐ C. 4.58×10^{3}

☐ D. 4.58×10^{5}

☐ E. 4.58×10^{6}

30) In a scatter plot showing the relationship between hours studied and exam scores, which of the following data points lies farthest from the line of best fit?

- A. (3, 75)
- B. (2, 22)
- C. (7, 70)
- D. (9, 90)
- E. No point

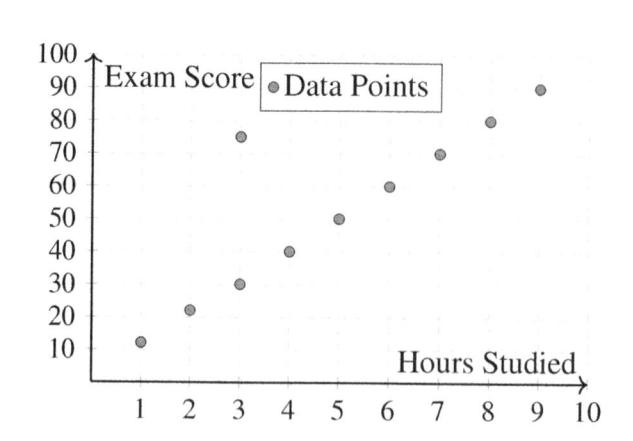

31) The graph of a quadratic function $h(x)$ was transformed to create the graph of $p(x) = h(x-2) - 3$. Which graph best represents p?

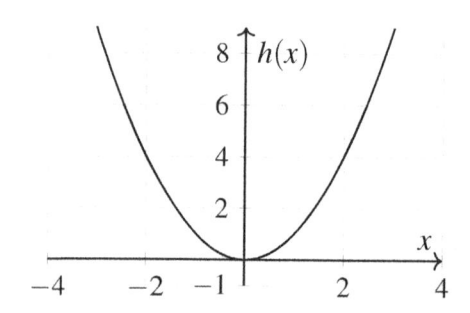

- A. 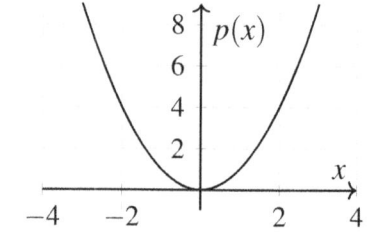
- B. 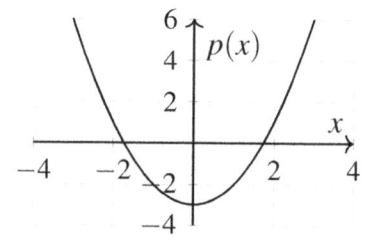
- C. 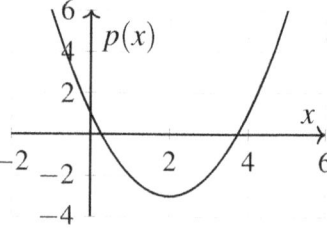
- D. 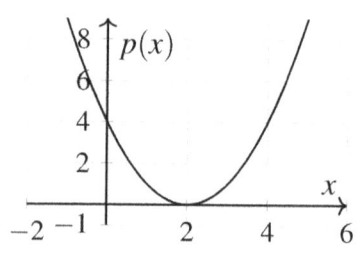
- E.

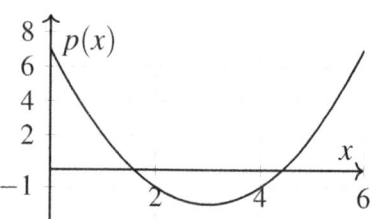

32) Solve for y in the equation: $(4)^2 + (-5)^2 + 3y - 12 = 32$.

2.1 Practices

- ☐ A. 1
- ☐ B. 2
- ☐ C. 3
- ☐ D. 4
- ☐ E. 5

33) A table of values for a quadratic function $h(x)$ is given as follows:

x	-4	-2	0
$h(x)$	0	-3	-4

What is the value of $h(4)$?

- ☐ A. -4
- ☐ B. 3
- ☐ C. 0
- ☐ D. 4
- ☐ E. -3

34) Consider the graph of the function $g(x) = 5(1.15)^x$. Which statement about this graph is true?

- ☐ A. The graph includes the point $(-2, 3)$.
- ☐ B. There is an asymptote to the equation of $y = 1$.
- ☐ C. The $x-$ intercept is 5.
- ☐ D. The graph includes the point $(0, 6)$.
- ☐ E. It is a increasing function.

35) A Venn diagram is used to represent the relationships among real numbers ($\mathbb{R}$), integers ($\mathbb{Z}$), and irrational numbers ($\mathbb{Q}'$). Which Venn diagram represents the relationships among these three sets of numbers?

- ☐ A. 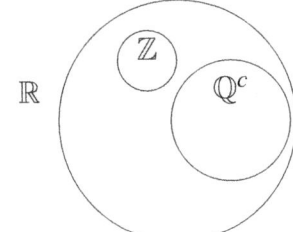
- ☐ B.

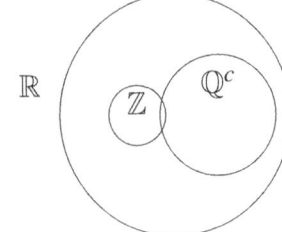

☐ C. ℝ ⊃ (ℤ, ℚᶜ overlapping)

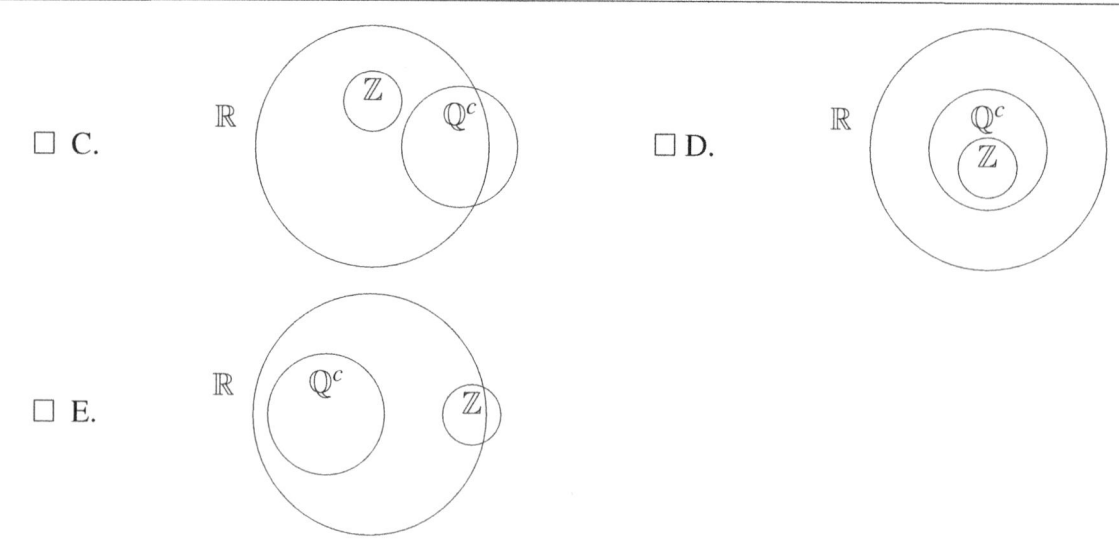

☐ D. ℝ ⊃ (ℚᶜ ⊃ ℤ)

☐ E. ℝ ⊃ (ℚᶜ), ℤ separate inside ℝ

36) Find the value of x that satisfies the equation $\frac{5}{x+2} = \frac{10}{24}$.

☐ A. 5

☐ B. 10

☐ C. 12

☐ D. 6

☐ E. 8

37) A school's weekly data has been plotted in a scatter plot, showing the relationship between the number of hours spent on extracurricular activities and students' average test scores. What can be inferred from the scatter plot?

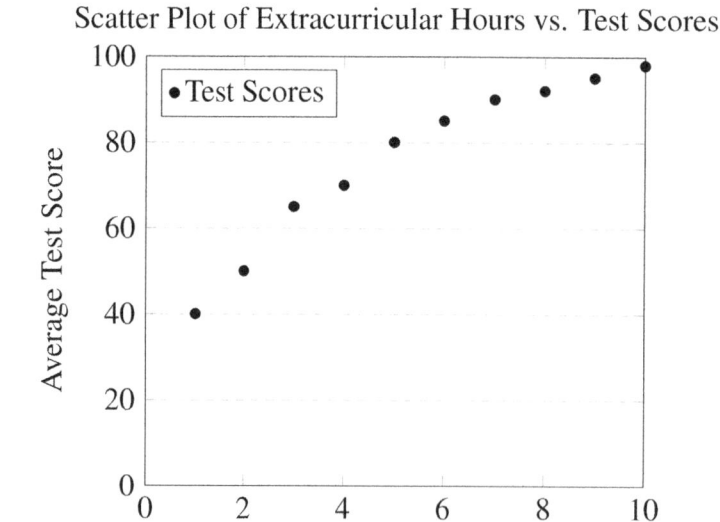

☐ A. Students' test scores increase as the hours spent on extracurricular activities increase.

2.1 Practices

☐ B. Students' test scores decrease as the hours spent on extracurricular activities increase.

☐ C. There is a consistent relationship between the hours spent on extracurricular activities and test scores.

☐ D. No clear pattern is observable, suggesting no correlation between extracurricular activities and test scores.

☐ E. The highest test scores are consistently reported mid-week regardless of the hours spent on extracurricular activities.

38) Two colleagues, Alex and Jordan, took loans to start their businesses. Alex borrowed $10,000 with a 6% annual simple interest rate for 4 years, while Jordan borrowed $10,000 with a 5.3% annual simple interest rate for 6 years. What is the difference in the total interest paid by Alex and Jordan?

☐ A. $380

☐ B. $420

☐ C. $520

☐ D. $680

☐ E. $780

39) In a summer camp, the number of participants on any day must be less than 30 for a field trip. If there are already 20 participants signed up, how many more participants can sign up?

☐ A. 9

☐ B. 10

☐ C. 11

☐ D. 12

☐ E. 13

40) Given the function $f(x) = \frac{1}{x+3} - 6$, for which value of x does $f(x) = -2$?

☐ A. $x = -5$

☐ B. $x = -\frac{5}{2}$

☐ C. $x = 0$

☐ D. $x = 2$

☐ E. $x = -\frac{11}{4}$

41) The function $g(x) = -(x+2)^2 + 3$ represents the height, in meters, of a ball thrown into the air after x seconds. What is the maximum height reached by the ball?

- A. 1 *m*
- B. 2 *m*
- C. 3 *m*
- D. 4 *m*
- E. 5 *m*

42) Simplify the function $g(x) = 3(1-2x)^2 - 9$ to its equivalent form.
- A. $g(x) = -12x^2 + 12x - 6$
- B. $g(x) = 12x^2 - 12x + 6$
- C. $g(x) = 12x^2 - 12x - 6$
- D. $g(x) = -12x^2 - 12x + 6$
- E. $g(x) = 12x^2 + 12x - 6$

43) On the number line shown below, which point best represents the value of $\sqrt{8}$?
- A. Point *A*
- B. Point *B*
- C. Point *C*
- D. Point *D*
- E. Point *E*

44) A diver's depth in meters below sea level after *t* seconds is recorded. The data is thought to fit a quadratic model.

Time, *t* (seconds)	Depth, *d(t)* (meters)
0	0
1	4
2	14
3	30
4	52

Which function best models the diver's depth over time?
- A. $d(t) = t^2 + 2t$
- B. $d(t) = 3t^2 + t$
- C. $d(t) = 4t^2 - t$

2.1 Practices

☐ D. $d(t) = 3t^2 + 3t + 1$

☐ E. $d(t) = 4t^2 + 4t + 1$

45) A square pyramid has a base side length of 6 meters and a height of 9 meters. What is its volume?

☐ A. 108 cubic meters

☐ B. 162 cubic meters

☐ C. 216 cubic meters

☐ D. 324 cubic meters

☐ E. 432 cubic meters

46) Simplify the following expression: $|8 - (-12) + (-6)| - |-20|$

☐ A. -18

☐ B. -16

☐ C. 6

☐ D. -6

☐ E. 18

47) Calculate the sum of $\sqrt{2x - 28}$ and $\sqrt{2x} - 14$ when $\sqrt{2x} = 8$.

☐ A. -6

☐ B. 2

☐ C. 8

☐ D. 14

☐ E. 0

48) A sequence can be generated by using $a_n = 2a_{n-1} + a_{n-2}$, where $a_1 = 4$, $a_2 = 1$ and n is a whole number greater than 1. What are the first five terms in the sequence?

☐ A. 4, 1, 2, 5, 12

☐ B. 4, 1, 6, 13, 32

☐ C. 4, 1, 6, 14, 33

☐ D. 4, 1, 5, 11, 24

☐ E. 4, 1, 9, 22, 49

49) Given $h(x) = x^2 - 6x - 16$, which statement is true?

☐ A. The zeroes are −4 and 4, because the factors of h are $(x-4)$ and $(x+4)$.

☐ B. The zeroes are 16 and −1, because the factors of h are $(x-16)$ and $(x+1)$.

☐ C. The zeroes are 8 and −2, because the factors of h are $(x-8)$ and $(x+2)$.

☐ D. The zeroes are 8 and 2, because the factors of h are $(x-8)$ and $(x-2)$.

☐ E. The zeroes are −8 and 2, because the factors of h are $(x+8)$ and $(x-2)$.

50) For what value(s) of x does the function $h(x) = \frac{2(x-1)}{(x-2)^2(x-1)}$ have a vertical asymptote?

☐ A. −2, 1

☐ B. 0, 2

☐ C. 1, 2

☐ D. 2

☐ E. 4

2.2 Answer Keys

1) B. $251.2\ cm^3$
2) D. Account Y would earn Lila about $58 more interest than Account X.
3) B. $30
4) D. -2
5) A. $(8+p, 6-q)$
6) C. $y = 3.00x + 2,000$
7) B. $V = \pi(5)^2(12)$
8) E. $8\ m, 15\ m, 17\ m$
9) C. $x = 3$
10) E.
11) A. 120
12) A. Domain: All real numbers; Range: $y \leq 5$
13) E. 35
14) B. -1
15) A. $\begin{cases} x - 2y = 12, \\ x + y = 3 \end{cases}$
16) B. $y = -\frac{1}{2}x^2 + 4$
17) C. 226 m^3
18) D. $\{(4,-1), (5,-2), (6,-3), (7,-4)\}$
19) D. $-\frac{1}{3}$
20) D. A line with a negative slope that does not pass through the origin.
21) A. A 2-year loan with a 9.5% annual simple interest rate.
22) E. $y = -200x + 9,000$
23) D. $4,186.67\ cm^3$
24) D. $5\ m$
25) A. $y = 0.5x$
26) D. Slope $= 1.5$, y-intercept $= 1$
27) D. $T = 2D + 80$
28) E. Screen E
29) E. 4.58×10^6
30) A. $(3, 75)$
31) C.
32) A. 1
33) C. 0
34) E. It is a increasing function.
35) A.
36) B. 10
37) A. Students' test scores increase as the hours spent on extracurricular activities increase.
38) E. $780
39) A. 9
40) E. $x = -\frac{11}{4}$
41) C. $3\ m$
42) C. $g(x) = 12x^2 - 12x - 6$
43) A. Point A
44) B. $d(t) = 3t^2 + t$
45) A. 108 cubic meters
46) D. -6
47) E. 0
48) B. 4, 1, 6, 13, 32
49) C. The zeroes are 8 and -2, because the factors of h are $(x-8)$ and $(x+2)$.
50) D. 2

2.3 Answers with Explanation

1) The volume V of a cone is given by the formula $V = \frac{1}{3}\pi r^2 h$, where r is the radius and h is the height. Plugging in the given values:

$$V = \frac{1}{3}\pi \times 4^2 \times 15 = \frac{1}{3}\pi \times 16 \times 15 = 80\pi = 80 \times 3.14 = 251.2\ cm^3.$$

Thus, the volume of the cone is $251.2\ cm^3$, which is option B.

2) For Account X (simple interest):

$$I_X = P \times r \times t = 3000 \times 0.028 \times 5 = \$420.$$

For Account Y (compound interest):

$$A_Y = P\left(1 + \frac{r}{n}\right)^{nt} = 3000\left(1 + \frac{0.03}{1}\right)^{1 \times 5} = 3000(1.03)^5 \approx \$3478.$$

Therefore, the interest earned for Account Y is:

$$I_Y = A_Y - P = \$3478 - \$3000 = \$478.$$

The difference in interest between Account Y and X is:

$$I_Y - I_X = \$478 - \$420 = \$58.$$

Therefore, account Y would earn Lila about $58 more interest than Account X, option D.

3) Since the cost is directly proportional to the hours, we can set up a proportion:

$$\frac{\text{Cost for 120 hours}}{120\ \text{hours}} = \frac{\text{Cost for 60 hours}}{60\ \text{hours}}$$

2.3 Answers with Explanation

Given that the cost for 120 hours is $60:

$$\frac{60}{120} = \frac{\text{Cost for 60 hours}}{60}$$

Simplifying, we find the cost for 60 hours:

$$\text{Cost for 60 hours} = \frac{60 \times 60}{120} = \$30.$$

Thus, the cost for running the machine for 60 hours is $30, which makes option B correct.

4) First, find the slope (m) of the line:

$$m = \frac{y_2 - y_1}{x_2 - x_1} = \frac{-5 - 3}{-3 - 5} = \frac{-8}{-8} = 1.$$

Use point-slope form, $y - y_1 = m(x - x_1)$, with point $(5, 3)$:

$$y - 3 = 1(x - 5)$$

Solving for the y-intercept (when $x = 0$):

$$y - 3 = 1(0 - 5), \Rightarrow y = 3 - 5 = -2.$$

Thus, the y-intercept of the function is -2, making option D correct.

5) The top-right vertex of the original rectangle is $(8, 6)$. After the translation, this point moves p units to the right and q units down. Therefore, the new coordinates of this vertex will be $(8 + p, 6 - q)$, option A.

6) To find the function, we use the given data:

$$3500 = 500m + b \quad \text{and} \quad 4400 = 800m + b.$$

Solving these equations simultaneously gives $m = 3$ (the rate per package) and $b = 2000$ (the fixed salary). Therefore, the total amount earned, y, if x packages are delivered, is given by $y = 3.00x + 2,000$, which is option C.

7) The volume of a cylinder is given by $V = \pi r^2 h$, where r is the radius and h is the height. Since the diameter is 10 inches, the radius r is half of that, which is 5 inches. Therefore, the volume V can be calculated using $V = \pi(5)^2(12)$, making option B correct.

8) For a right triangle, the Pythagorean theorem must hold, where the square of the hypotenuse (the longest side) equals the sum of the squares of the other two sides. Testing the options, option E satisfies this condition: $8^2 + 15^2 = 64 + 225 = 289$, which is equal to 17^2. Hence, the side lengths of 8 m, 15 m, and 17 m can form a right triangle.

9) The axis of symmetry of a quadratic function graph is a vertical line that passes through the vertex of the parabola. In this case, $x = 3$ would be an axis of symmetry as it is a vertical line that can intersect the vertex of a typical quadratic graph. The other options either represent horizontal lines or are less likely to intersect the vertex based on typical quadratic shapes. Therefore, the correct answer is option C.

10) To determine which graph represents the solution set of the inequality $x > 2y - 5$, let's first rearrange it into a more familiar form:

$$x > 2y - 5 \Rightarrow y < \frac{x}{2} + \frac{5}{2}.$$

This inequality describes a region in the xy-plane. The line $y = \frac{x}{2} + \frac{5}{2}$ serves as the boundary. Since it's a "less than" inequality, the solution set is the area below this line.

Therefore, the correct option is E. This graph properly illustrates the inequality $x > 2y - 5$ by shading the region below the line $y = \frac{x}{2} + \frac{5}{2}$.

11) Based on the table provided, the relationship between the number of books in stock and the number of daily sales appears to be linear. We can observe that for each increase of 1,000 books in stock, daily sales increase by 15 (from 60 to 75, 75 to 90, and 90 to 105 respectively).

To predict the daily sales at 5,000 books in stock, we extend this linear pattern. Given that sales increase by 15 for every additional 1,000 books, an inventory of 5,000 books (which is 1,000 more than the last data point at 4,000 books) would likely result in an increase of 15 sales from the 105 sales at 4,000 books.

Thus, the predicted number of daily sales for 5,000 books in stock is:

$$105 \text{ sales} + 15 \text{ sales} = 120 \text{ sales}.$$

2.3 Answers with Explanation

Therefore, the best prediction for the number of daily sales at 5,000 books in stock is 120 sales.

12) The domain of a quadratic function is always all real numbers because there are no restrictions on the input values (*x* values). Since the graph opens downwards and the vertex is the highest point at $(3,5)$, the range is $y \leq 5$, as all output values (*y* values) are less than or equal to 5. Therefore, option A is correct.

13) In a parallelogram, adjacent angles are supplementary. Therefore, we have the equation:

$$(2x+10)° + (3x-5)° = 180°.$$

Simplifying and solving for *x*, we get:

$$5x + 5 = 180 \Rightarrow 5x = 175 \Rightarrow x = 35.$$

Thus, the value of *x* is 35, which is option E.

14) The rate of change, or slope, of a line is calculated as the change in *y* divided by the change in *x*. For the points $(1,4)$ and $(4,1)$, the slope is:

$$\text{Slope} = \frac{\Delta y}{\Delta x} = \frac{1-4}{4-1} = \frac{-3}{3} = -1.$$

Therefore, the rate of change of the line is -1, which is option B.

15) Using the points of each line, we can determine the equations. For L_1, using $(2,-5)$ and $(4,-4)$, the slope is $\frac{-4+5}{4-2} = \frac{1}{2}$. The equation is $y = \frac{1}{2}x - 6$. For L_2, using $(-4,7)$ and $(0,3)$, the slope is $\frac{3-7}{0+4} = -1$. The equation is $y = -x + 3$. Rewriting in standard form:

$$L_1 : x - 2y = 6; \quad L_2 : x + y = 3.$$

Thus, the system is $\begin{cases} x - 2y = 12, \\ x + y = 3 \end{cases}$, which is option A.

16) To determine which equation best describes the given graph of a quadratic function, we analyze the characteristics of the graph:

- The graph opens downwards, indicating a negative coefficient for the x^2 term. This eliminates options with a positive coefficient for x^2.
- The vertex of the parabola is at its maximum point with a y-coordinate of 4. Therefore, the constant term in the equation should be 4.
- The width of the parabola is determined by the coefficient of x^2. A smaller absolute value of this coefficient suggests a wider parabola. The graph shown has a relatively wide parabola.

Given these observations, the equation that best matches the graph is:

$$y = -\frac{1}{2}x^2 + 4$$

This equation has:

- A negative coefficient for the x^2 term, indicating a downward opening.
- A constant term of 4, matching the y-coordinate of the vertex.
- A relatively small absolute coefficient for x^2, indicating a wider parabola.

Therefore, the correct option is B.

17) The volume of a cone is given by $V = \frac{1}{3}\pi r^2 h$, where r is the radius and h is the height. For this hill, $r = \frac{12}{2} = 6$ meters and $h = 6$ meters. Calculating the volume:

$$V = \frac{1}{3}\pi(6)^2(6) = \frac{1}{3} \times 3.14 \times 36 \times 6 \approx 226 \text{ m}^3.$$

Therefore, the closest volume measurement is 226 m^3, which is option C.

18) A function must have a unique y value for each x value. In option D, each x value is paired with one unique y value. The other options have at least one x value that corresponds to multiple y values, violating the definition of a function. Thus, option D correctly represents a function.

19) The zero of a linear function is the x-coordinate where the graph intersects the x-axis. The graph passes through two points $(0, 1)$ and $(-1, -2)$. Thus, the slpoe is:

$$\text{Slope} = \frac{1-(-2)}{0-(-1)} = 3.$$

2.3 Answers with Explanation

Therefore, the equation of the line is $y = 3x + 1$. To find the x-intercept, we set $y = 0$ and get:

$$0 = 3x + 1 \Rightarrow 3x = -1 \Rightarrow x = -\frac{1}{3}.$$

Therefore, the zero of the function is at $x = -\frac{1}{3}$, corresponding to option D.

20) A non-proportional linear relationship is represented by a linear function that does not pass through the origin. Options A and E represent proportional relationships as they pass through the origin. Option B represents a constant function, and option C is not a function. Therefore, option D, representing a line with a negative slope that does not pass through the origin, correctly illustrates a non-proportional linear relationship.

21) To find the total interest paid, calculate the interest for each option:
- For option A: $10,000 \times 9.5\% \times 2 = \$1,900$ total interest.
- For option B: $10,000 \times 7.5\% \times 3 = \$2,250$ total interest.
- For option C: $10,000 \times 5.8\% \times 4 = \$2,320$ total interest.
- For option D: $10,000 \times 5.0\% \times 5 = \$2,500$ total interest.
- For option E: $10,000 \times 4.8\% \times 6 = \$2,880$ total interest.

The least total interest is paid with option A.

22) The equation should represent a depreciation in value over time. The negative coefficient before x indicates depreciation. The initial value of the computer at $x = 0$ should be the y-intercept, which is $9,000$.

Assuming a linear depreciation, an equation with a format $y = mx + b$, where m is negative and b is the initial value, is needed. Option E, with $y = -200x + 9,000$, fits these criteria, representing a yearly depreciation of 200 dollars from an initial value of $9,000$.

23) To find the volume of a sphere, use the formula $V = \frac{4}{3}\pi r^3$, where r is the radius. The radius is half the diameter, so for a diameter of $20cm$, the radius is $10cm$. Thus, the volume is:

$$V = \frac{4}{3}\pi(10)^3 = \frac{4}{3} \times \pi \times 1000 \approx 4,186.67 \ cm^3.$$

Therefore, the volume of the globe is approximately $4,186.67 \ cm^3$, which is option D.

24) For a rectangular prism, surface area S is given by $S = 2lw + 2lh + 2wh$. Here, $l = 2 \ m$, $w = 5 \ m$, and

$S = 90 \, m^2$. Rearranging the formula for height h gives:

$$90 = 2(2 \times 5) + 2(2h) + 2(5h).$$

Simplifying and solving for h:

$$90 = 20 + 4h + 10h \Rightarrow 14h = 70 \Rightarrow h = \frac{70}{14} = 5 \, m.$$

The correct answer is $5 \, m$, option D.

25) The rate of travel is calculated by dividing the distance by time. Thus:

$$\text{Rate} = \frac{104 \text{ km}}{208 \text{ minutes}} = 0.5 \text{ km/minute}.$$

The option A, $y = 0.5x$, represents a rate of 0.5 or half the distance per time unit, aligning with 0.5 km/minute.

26) The slope of a line is calculated by the change in y divided by the change in x (rise over run). For points $(2,4)$ and $(-2,-2)$:

$$\text{Slope} = \frac{4-(-2)}{2-(-2)} = \frac{6}{4} = 1.5.$$

Thus, the equation of the line is $y = 1.5x + 1$. The y-intercept is the y-value where the line crosses the y-axis, which we can get by putting x equal to zero in the equation of the line:

$$x = 0 \Rightarrow y = 1.5 \times 0 + 1 \Rightarrow y = 1.$$

Therefore, the slope is 1.5 and the y-intercept is 1, which corresponds to option D.

27) The total water in the tank increases by 2 liters each day. The function representing this situation is linear, starting with 80 liters and increasing by 2 liters daily. Therefore, the equation is:

$$T = 2D + 80.$$

Where T is the total water in liters and D is the number of days. Option D correctly represents this relationship.

2.3 Answers with Explanation 31

28) Comparing the diagonal lengths in inches:

- Screen A: $\frac{15}{4} = 3.75$
- Screen B: $\frac{10}{3} \approx 3.33$
- Screen C: $\sqrt{5} \approx 2.24$
- Screen D: $\sqrt{20} \approx 4.47$
- Screen E: $\frac{\sqrt{16}}{4} = \frac{4}{4} = 1$

The shortest diagonal length is for Screen E, which is 1 inch. Therefore, option E is correct.

29) To write a number in scientific notation, you move the decimal point to the left until only one digit remains to the left. The number of places moved gives the exponent on 10:

$$4,580,000 = 4.58 \times 10^6.$$

Thus, the population in scientific notation is 4.58×10^6, making option E correct.

30) This question requires the line of best fit to be drawn on the scatter plot, which is the line $y = 10x$. The data point farthest from this line will be the answer, which is $(3, 75)$, option A.

31) The original graph of $h(x)$ is a standard upward-opening parabola. Among the options, Option C shows the correct transformation of shifting the graph 2 units to the right and 3 units downward, which matches $p(x) = h(x-2) - 3$.

32) Simplify the given equation:
$$16 + 25 + 3y - 12 = 32,$$

Combine like terms:
$$29 + 3y = 32,$$

Solve for y:
$$3y = 32 - 29 \Rightarrow 3y = 3 \Rightarrow y = 1.$$

Thus, the value of y is 1, making option A correct.

33) Consider h as $h(x) = ax^2 + bx + c$. From the table, we have:

$$h(0) = -4 \Rightarrow c = -4,$$

$$h(-4) = 0 \Rightarrow 16a - 4b = 4 \Rightarrow 4a - b = 1,$$

$$h(-2) = -3 \Rightarrow 4a - 2b = 1.$$

Solving the system of equation

$$\begin{cases} 4a - b = 1, \\ 4a - 2b = 1 \end{cases}$$

we get: $b = 0$ and $a = \frac{1}{4}$. Thus,

$$h(x) = \frac{1}{4}x^2 - 4.$$

Therefore, $h(4) = \frac{1}{4} \times 16 - 4 = 0$, which is option C.

34) The function $g(x) = 5(1.15)^x$ is an exponential function. The base of the exponential, 1.15, is greater than 1, which indicates that the function is increasing. None of the other options (A, B, C, D) which imply a point on the graph or a specific intercept are correct. The correct statement is that the function is increasing, making option E the correct choice.

35) Real numbers include all possible numbers along the number line. An integer may be regarded as a real number that can be written without a fractional component. While Rational numbers are those that can be expressed as a fraction of two integers, and Irrational numbers cannot be expressed this way. The two subsets (Rational and Irrational) do not overlap but are both entirely contained within the set of Real numbers. Every integer is also a rational number. Thus, the set of integers is a subset of rational numbers and hence there is no overlap between integers and irrational numbers. Therefore, the option A is the correct Venn diagram.

36) To find the value of x that satisfies the given equation, cross multiply to get:

$$5 \times 24 = 10 \times (x + 2),$$

2.3 Answers with Explanation

Simplify the equation and solve for x:

$$120 = 10x + 20 \Rightarrow 10x = 120 - 20 \Rightarrow 10x = 100 \Rightarrow x = 10.$$

Therefore, the correct answer is B.

37) The scatter plot suggests a positive correlation between the hours spent on extracurricular activities and students' test scores. As the number of hours increases, there is a general trend of improvement in test scores, indicating that increased engagement in extracurricular activities might be associated with better academic performance.

38) Calculate each interest by formula $I = P \cdot r \cdot t$, we get:

$$\text{Alex's interest: } \$10,000 \times 6\% \times 4 = \$2400,$$

$$\text{Jordan's interest: } \$10,000 \times 5.3\% \times 6 = \$3180,$$

$$\text{Difference: } \$3180 - \$2400 = \$780.$$

Therefore, the difference in interest paid is $780, which is option E.

39) The inequality representing the situation is $20 + x < 30$.

Solving for x, we get $x < 10$.

Since x must be a positive integer, the maximum number of additional participants is 9, which is option A.

40) To find the value of x that makes $f(x) = -2$, set the function equal to -2 and solve for x:

$$\frac{1}{x+3} - 6 = -2 \Rightarrow \frac{1}{x+3} = 4 \Rightarrow 1 = 4(x+3) \Rightarrow 4x + 12 = 1 \Rightarrow x = -\frac{11}{4}.$$

Therefore, the value of x is $x = -\frac{11}{4}$, which corresponds to option E.

41) The vertex form of the quadratic function gives the maximum height directly as the y-coordinate of the vertex:

$$g(x) = -(x+2)^2 + 3$$

The maximum height, which is the y-value of the vertex, is 3 m. Therefore, the correct option is C.

42) To simplify the function, expand the squared term and distribute the 3, then subtract 9:

$$g(x) = 3(1 - 4x + 4x^2) - 9 = 3 - 12x + 12x^2 - 9 = 12x^2 - 12x - 6.$$

Therefore, the function simplifies to $12x^2 - 12x - 6$, which corresponds to option C.

43) The value of $\sqrt{8}$ is approximately 2.828. On the number line, this value is closest to point A. Therefore, the best representation for the value of $\sqrt{8}$ is point A.

44) To determine the best model, we can use the data points to find a quadratic equation of the form $d(t) = at^2 + bt + c$. Using the points $(0,0)$, $(1,4)$, and $(2,14)$:

At $t = 0$, $d(0) = 0$, which implies $c = 0$. At $t = 1$, $d(1) = 4$, which gives $a + b = 4$. At $t = 2$, $d(2) = 14$, which gives $4a + 2b = 14$.

Solving for a and b, we find that $a = 3$ and $b = 1$. Thus, the model is:

$$d(t) = 3t^2 + t.$$

This corresponds to option B.

45) Using the volume formula $V = \frac{1}{3}x^2 h$, with $x = 6$ meters and $h = 9$ meters, we find:

$$V = \frac{1}{3}(6)^2(9) = \frac{1}{3}(36)(9) = \frac{1}{3}(324) = 108 \text{ cubic meters.}$$

Therefore, the volume of the square pyramid is 108 cubic meters, which corresponds to option A.

46) First, simplify the expression inside the absolute value:

$$|8 - (-12) + (-6)| = |8 + 12 - 6| = |14| = 14,$$

Then, evaluate the second absolute value:

$$|-20| = 20,$$

2.3 Answers with Explanation

Finally, subtract the two absolute values:

$$14 - 20 = -6.$$

Therefore, the correct answer is option D.

47) Given that $\sqrt{2x} = 8$, we can square both sides to find $2x$:

$$(\sqrt{2x})^2 = 8^2 \Rightarrow 2x = 64 \Rightarrow x = 32,$$

Substitute x into the original expressions:

$$\sqrt{2(32) - 28} + \sqrt{2(32)} - 14 = \sqrt{64 - 28} + 8 - 14 = \sqrt{36} - 6 = 0.$$

Therefore, the correct answer is option E.

48) Using the recursive formula, calculate the next terms:

$$a_3 = 2a_2 + a_1 = 2(1) + 4 = 6,$$

$$a_4 = 2a_3 + a_2 = 2(6) + 1 = 13,$$

$$a_5 = 2a_4 + a_3 = 2(13) + 6 = 32.$$

Therefore, the first five terms of the sequence are 4, 1, 6, 13, and 32, making option B correct.

49) To factor the quadratic equation $h(x) = x^2 - 6x - 16$, we need two numbers that thier sum is -6 and their product is -16, which are -8 and 2. Thus, we get:

$$h(x) = x^2 - 6x - 16 = (x - 8)(x + 2).$$

Therefore, the zeroes of $h(x)$ are $x = 8$ and $x = -2$.

50) The vertical asymptote of $h(x)$ occurs at the value(s) of x where the denominator equals zero, and the factor does not cancel out with any factor in the numerator. After canceling out the common factor of $(x - 1)$, the function $h(x)$ has a vertical asymptote at $x = 2$, since this is the point where the remaining denominator $(x - 2)^2$ equals zero.

3. Practice Test 2

CBEST Math Practice Test

Total number of questions: 50

Total time: 90 Minutes

Calculators are prohibited for the CBEST exam.

3.1 Practices

1) Identify the set of ordered pairs that define y as a function of x.
   - ☐ A. $\{(3,2), (3,-2), (4,5), (4,-5)\}$
   - ☐ B. $\{(-1,0), (0,1), (1,2), (2,3)\}$
   - ☐ C. $\{(1,3), (2,2), (3,3), (2,3)\}$
   - ☐ D. $\{(6.5,2), (6.5,-2), (7.5,4), (7.5,-4)\}$
   - ☐ E. $\{(-2.4,7), (-1.4,6), (0.6,5), (0.6,4)\}$

3.1 Practices

2) Determine the slope of the line that passes through the points $(2, -3)$ and $(7, 12)$.

- ☐ A. $\frac{3}{5}$
- ☐ B. 3
- ☐ C. $\frac{1}{3}$
- ☐ D. 1
- ☐ E. $\frac{2}{15}$

3) Consider a rectangular prism with a height of 8 inches, a width of 5 inches, and a length of 7 inches. Calculate the total surface area of this prism in square inches.

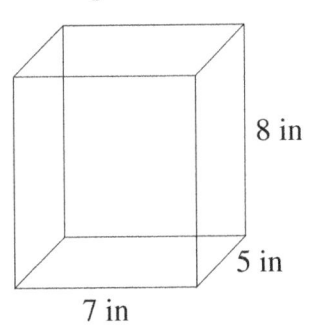

8 in
5 in
7 in

- ☐ A. 142
- ☐ B. 214
- ☐ C. 286
- ☐ D. 262
- ☐ E. 430

4) Emily allocates a total of $1500 between two separate savings accounts. The first account, X, offers a yearly simple interest rate of 4%. The second account, Y, provides an annual compound interest rate of 4%. Without any additional deposits or withdrawals, determine the difference between the amount of interest accumulated by both Account X and Account Y after a duration of 4 years.

- ☐ A. $2.40
- ☐ B. $9.60
- ☐ C. $24.00
- ☐ D. $14.78
- ☐ E. $48.00

5) Consider the following graph of the function $g(x) = 6x^2 + 4x - 5$:

$$y = 6x^2 + 4x - 5$$

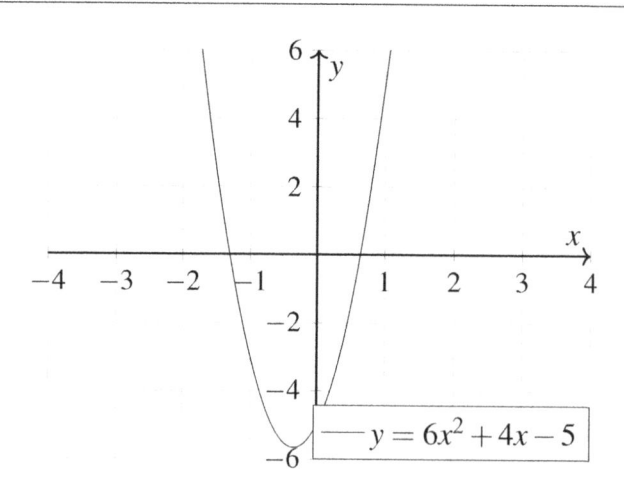

Find the approximate zeros of g?

☐ A. −1 and 5

☐ B. $\frac{5}{3}$ and $\frac{1}{2}$

☐ C. $-\frac{4}{3}$ and $\frac{2}{3}$

☐ D. $-\frac{1}{2}$

☐ E. $\frac{5}{6}$ and $\frac{1}{3}$

6) On the grid shown below, there are five different places marked. Identify which place on the grid is located at $(3, -2)$.

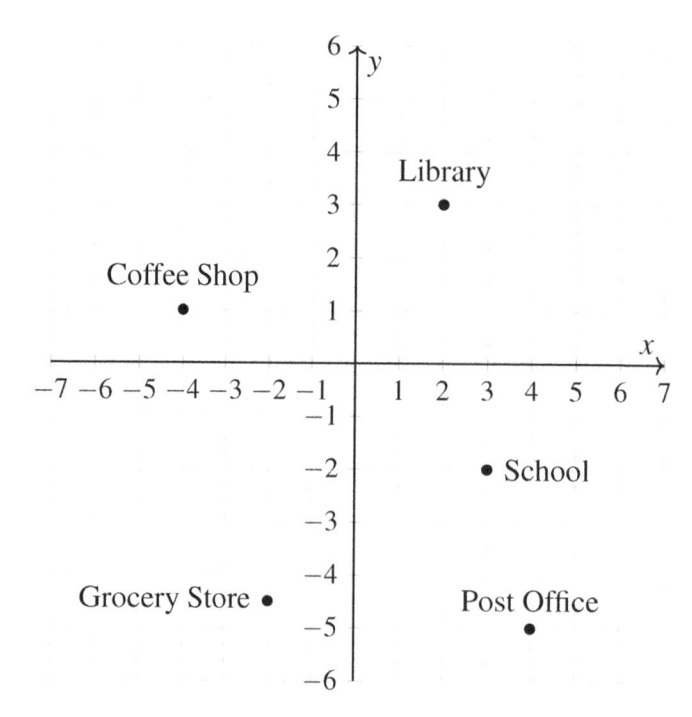

☐ A. Library

3.1 Practices

☐ B. Coffee Shop

☐ C. Grocery Store

☐ D. School

☐ E. Post Office

7) Given the inequality $x > -1$, what is a possible value for the expression $x+2$ in the following equation?

$$x+2 = \frac{7(x+1)}{(x^2+3x+2)}.$$

☐ A. -5

☐ B. -2

☐ C. 4

☐ D. $\sqrt{7}$

☐ E. 16

8) If the measure of a right angle is given as $(4x-10)°$, what could be a value of x?

☐ A. 25

☐ B. 32.5

☐ C. 40.2

☐ D. 55

☐ E. 65

9) Which graph corresponds to $y = \frac{1}{2}x - 2$?

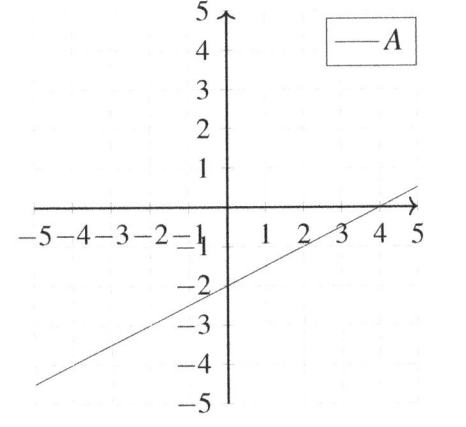

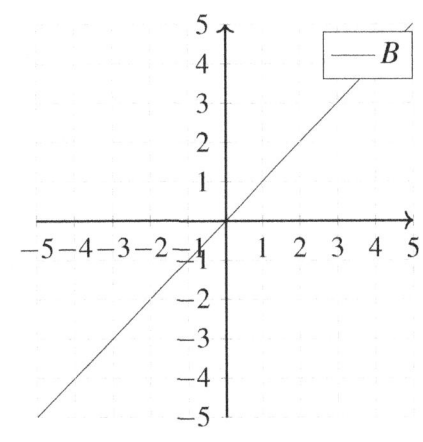

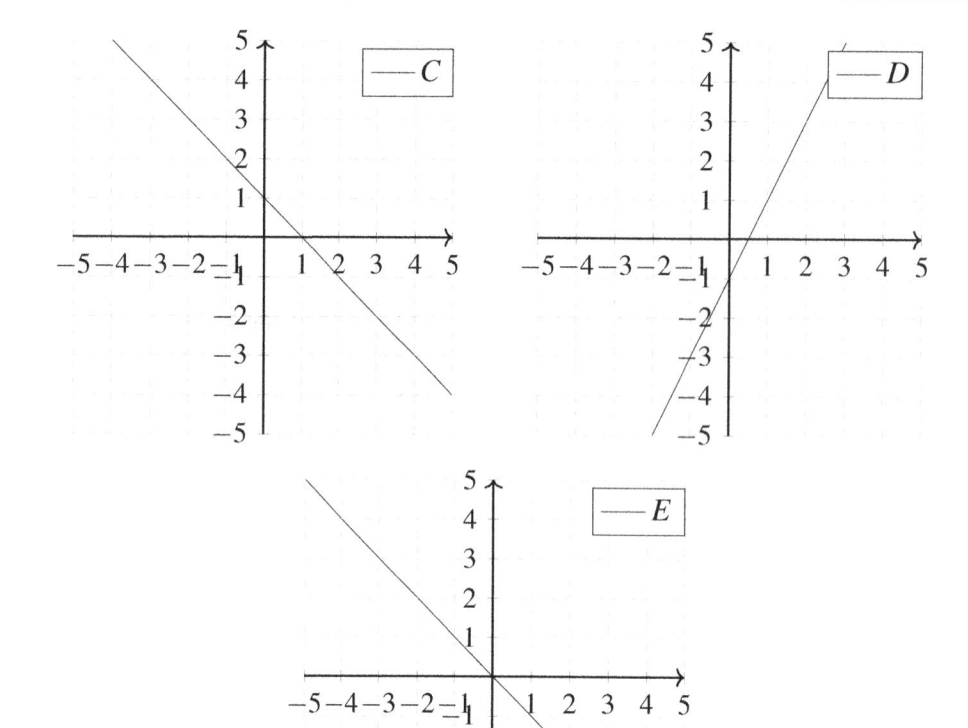

☐ A. Graph A

☐ B. Graph B

☐ C. Graph C

☐ D. Graph D

☐ D. Graph E

10) Which sets of measurements could correspond to the sides of a right triangle in centimeters?

☐ A. 2 cm, 3 cm, 6 cm

☐ B. 4 cm, 5 cm, 6 cm

☐ C. 5 cm, 12 cm, 13 cm

☐ D. 7 cm, 8 cm, 9 cm

☐ E. 8 cm, 9 cm, 11 cm

11) Calculate the sum of the following mixed numbers: $2\frac{3}{5} + 1\frac{1}{4} + 4\frac{2}{20} + 3\frac{1}{2}$

☐ A. $12\frac{1}{2}$

3.1 Practices

- [ ] B. $11\frac{3}{4}$
- [ ] C. $11\frac{5}{6}$
- [ ] D. 12
- [ ] E. $11\frac{9}{20}$

12) Two online streaming services offer different subscription plans. Service A charges a monthly fee of $10.00 and $0.50 per movie watched. Service B charges a monthly fee of $5.00 and $1.00 per movie. After how many movies will the total cost for both services be the same?

- [ ] A. 5
- [ ] B. 10
- [ ] C. 15
- [ ] D. 20
- [ ] E. 25

13) Two classes attended a science exhibition.

- Class A purchased 10 tickets with a $50 group discount.

- Class B purchased 5 tickets with a $25 group discount.

- Both classes spent the same amount in total.

What was the price of each individual ticket?

- [ ] A. $5
- [ ] B. $10
- [ ] C. $15
- [ ] D. $20
- [ ] E. $25

14) Two lines on a coordinate grid each represent an equations. Identify the ordered pair that simultaneously satisfies both of the equations.

- A. $(-2, 1)$
- B. $(-1, -1)$
- C. $(1, 10)$
- D. $(2, 13)$
- E. $(3, 16)$

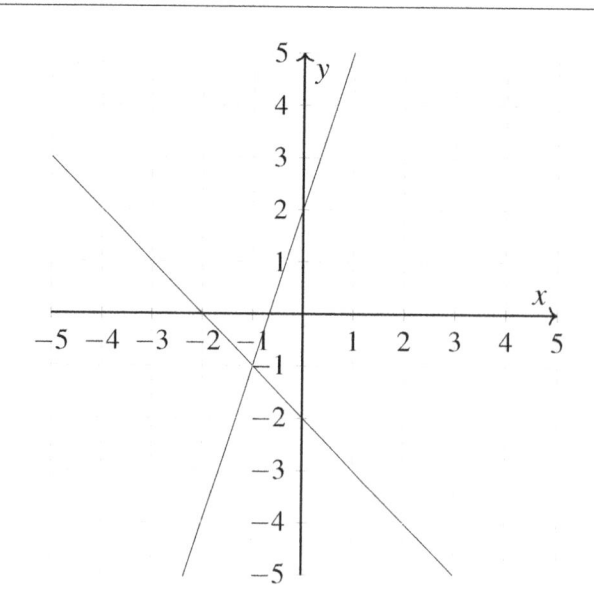

15) Quadrilateral $DEFG$ is rotated $180°$ about the origin to form quadrilateral $D'E'F'G'$.

Which statement is NOT true?

- A. The sum of the angle measures of quadrilateral $DEFG$ is equal to the sum of the angle measures of quadrilateral $D'E'F'G'$.
- B. The angle measures of quadrilateral $DEFG$ are greater than the corresponding angle measures of quadrilateral $D'E'F'G'$.
- C. Quadrilateral $DEFG$ is congruent to quadrilateral $D'E'F'G'$.
- D. The area of quadrilateral $DEFG$ is equal to the area of quadrilateral $D'E'F'G'$.
- E. The perimeter of quadrilateral $DEFG$ is equal to the perimeter of quadrilateral $D'E'F'G'$.

16) Which scenario is best described by the equation $4x + 30 = 8x$?

- A. Tom can bake 4 cakes per day, while Jerry can bake 8 cakes per day. How many days, x, would it take for Tom and Jerry to have baked the same number of cakes?
- B. Tom paid a deposit of $30 for a gym membership, plus $8 per session. Jerry paid $4 per session. How many sessions, x, would it take for Tom and Jerry to spend the same amount?

3.1 Practices

☐ C. Tom can swim 4 laps in an hour, while Jerry can swim 8 laps per hour. Tom already swam 30 laps. How many hours, x, would it take for Tom and Jerry to have swum the same number of laps?

☐ D. Tom earns $8 per hour working part-time. Jerry earns $30 per hour. How many hours, x, would it take for Tom to earn the same amount as Jerry?

☐ E. Tom is cycling at $30\frac{km}{h}$, while Jerry is cycling at $8\frac{km}{h}$. How long will it take for Tom to overtake Jerry?

17) The table below displays selected points from the graph of a linear function denoted by h.

x	0	3	5	7
$h(x)$	3	6	8	10

Following a horizontal translation of the graph of h by 4 units to the right, a new graph representing the function j is formed. What is the accurate comparison between the graphs of h and j?

☐ A. The x-intercept of the graph of h is 4 units below the x-intercept of the graph of j.

☐ B. The graph of h is steeper than the graph of j.

☐ C. The y-intercept of the graph of h is 12 units to the right of the y-intercept of the graph of j.

☐ D. The graph of h is less steep than the graph of j.

☐ E. The y-intercept of the graph of h is 4 units above of the y-intercept of the graph of j.

18) Consider a playground slide shaped like a right triangle. Given the lengths of two sides in meters, determine the length of the slide, in meters.

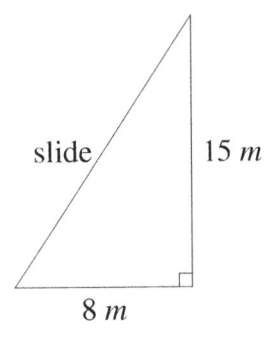

slide, 15 m, 8 m

What is the length of the slide in meters?

☐ A. 17

☐ B. 19

☐ C. 23

☐ D. 25

☐ E. 27

19) At a book fair, the cost of 5 pencils is 10, with each pencil having the same price. Express this situation as an equation representing the cost of each pencil at the fair.

☐ A. $5p = 1$

☐ B. $5p = 10$

☐ C. $5p + 10 = 0$

☐ D. $10p = 1$

☐ E. $10p = 5$

20) Consider the following quadratic functions:

$$f(x) = x^2 + 4,$$
$$g(x) = -2x^2 + 5,$$
$$h(x) = 3x^2 - 6.$$

Which statement about these functions is false?

☐ A. Two of these functions' graphs have a maximum point.

☐ B. All these functions' graphs have the same axis of symmetry.

☐ C. Two of these functions' graphs cross the x-axis.

☐ D. All these functions' graphs have different y-intercepts.

☐ E. Functions $g(x)$ and $h(x)$ intersect each other.

21) The linear functions $h(x)$ and $k(x)$ are graphed in the coordinate plane. How was the graph of h altered to produce the graph of k?

3.1 Practices

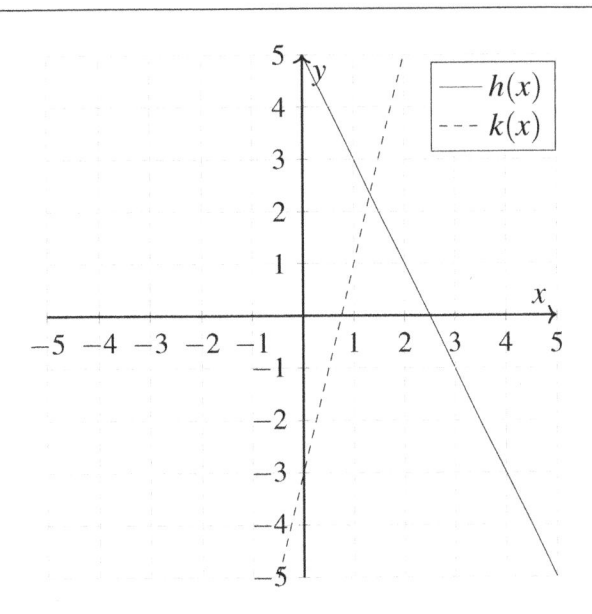

☐ A. The slope was multiplied by -2, and the y-intercept was decreased by 8.

☐ B. The slope was multiplied by 2, and the y-intercept was increased by 8.

☐ C. The slope was multiplied by 2, and the y-intercept was decreased by 8.

☐ D. The slope was multiplied by -2, and the y-intercept was decreased by 2.

☐ E. The slope was multiplied by -2, and the y-intercept was increased by 2.

22) Mark is a graphic artist. Each month he earns a fixed salary plus extra money for each project he completes.

- In March, Mark finished 35 projects and received a total payment of $1800.

- In April, he finished 70 projects and his total payment was $2500.

Determine the function that calculates y, Mark's total monthly earnings, given that he completes x number of projects.

☐ A. $y = 20x$

☐ B. $y = 20x + 1100$

☐ C. $y = 70x + 1800$

☐ D. $y = 100x$

☐ E. $y = 100x + 20$

23) A study suggests that as the amount of weekly physical activity increases for adults, their reported stress levels decrease. Which scatterplot could support the study's findings?

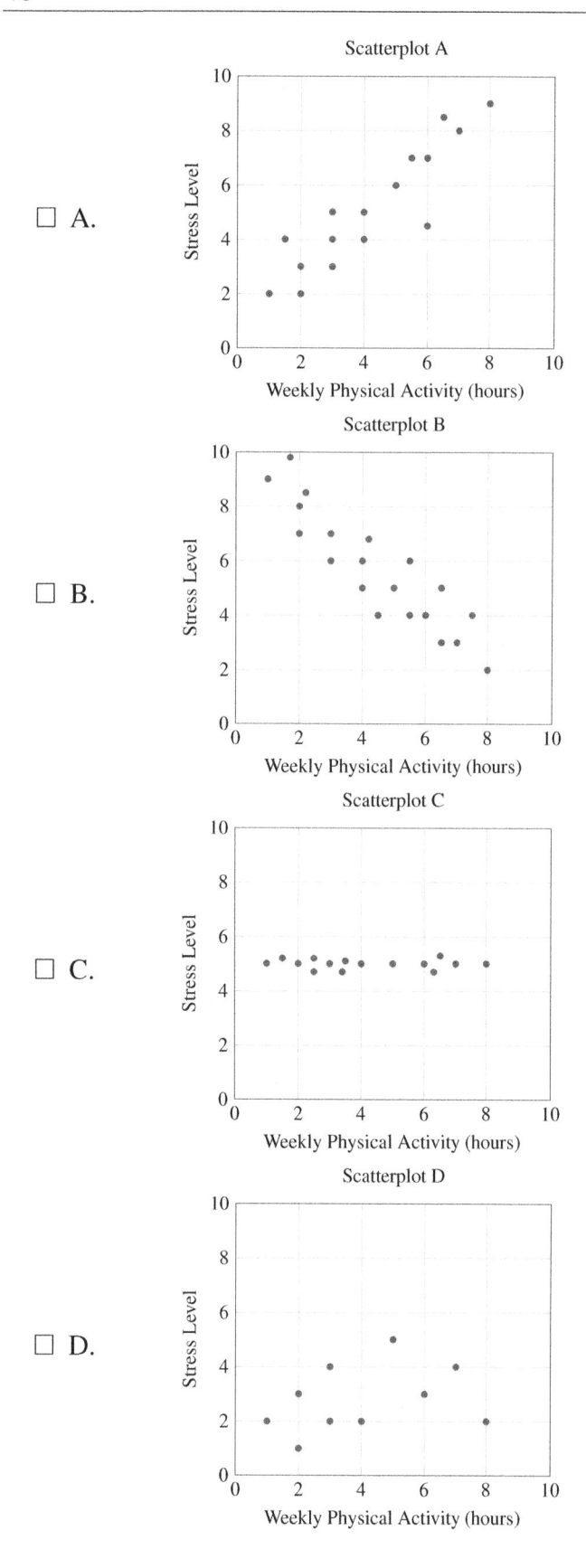

☐ A.

Scatterplot A

☐ B.

Scatterplot B

☐ C.

Scatterplot C

☐ D.

Scatterplot D

3.1 Practices

☐ E.

Scatterplot E

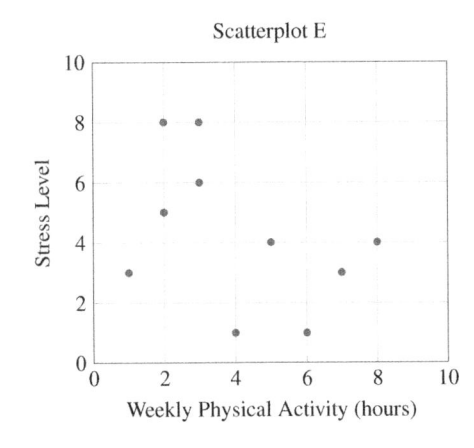

24) Linda invested $3000 into a savings account with an annual simple interest rate of 4%. Without any additional deposits or withdrawals, determine the amount of interest earned after 5 years.

☐ A. $60
☐ B. $600
☐ C. $1200
☐ D. $6000
☐ E. $12000

25) Consider the initial terms of a sequence as follows: 9, 13, 18, 24, 31, ⋯. Which formula is suitable to find the nth term of this sequence?

☐ A. $\frac{1}{2}n(n+5)$
☐ B. $\frac{1}{2}n(n+5)+6$
☐ C. $\frac{1}{2}n(n+5)-5$
☐ D. $4n+5$
☐ E. $5n+4$

26) Examine the graph of the function g plotted in the Cartesian plane. Identify the x-coordinate where $g(x)$ reaches its minimum.

- [ ] A. −4
- [ ] B. −1
- [ ] C. 1
- [ ] D. 3
- [ ] E. 6

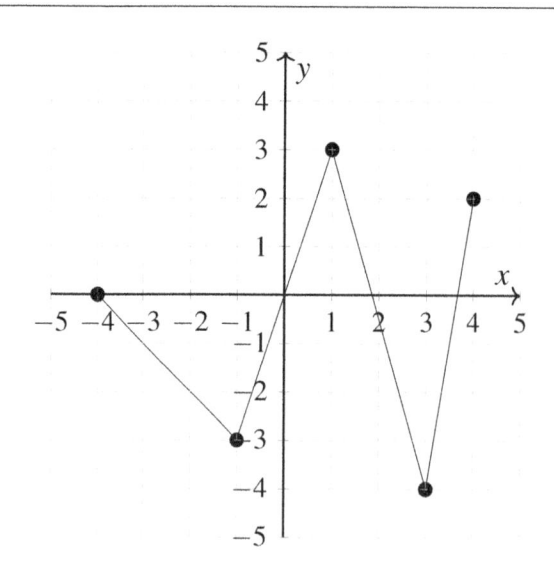

27) A metallic ball used in a machine part has a diameter of 18*cm*. Calculate its volume in cubic centimeters. Consider $\pi = 3.14$.

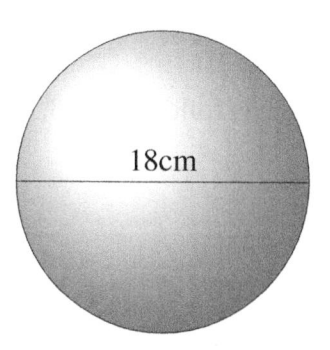

- [ ] A. $3052.08 cm^3$
- [ ] B. $6104.2 cm^3$
- [ ] C. $12208.5 cm^3$
- [ ] D. $24416.9 cm^3$
- [ ] E. $48833.8 cm^3$

28) Find the equation of the line that goes through the points $(3,5)$ and $(-2,-4)$.

3.1 Practices

☐ A. $y = \frac{9}{5}x + \frac{4}{5}$

☐ B. $y = \frac{9}{5}x - \frac{2}{5}$

☐ C. $y = -\frac{9}{5}x + \frac{4}{5}$

☐ D. $y = \frac{5}{9}x - \frac{14}{9}$

☐ E. $y = -\frac{5}{9}x + \frac{14}{9}$

29) In an art class, students are working with either clay or paint. The table details the distribution in two groups. What is the proportion of Group 2 students using only paint?

	Clay	paint	Total
Group 1	18	8	26
Group 2	15	20	35
Total	33	28	61

☐ A. $\frac{8}{35}$

☐ B. $\frac{15}{26}$

☐ C. $\frac{18}{35}$

☐ D. $\frac{20}{35}$

☐ E. $\frac{23}{26}$

30) Determine the value of x that makes the equation $5x - 8 = 7x + 4$ true.

☐ A. -6

☐ B. -3

☐ C. 0

☐ D. 3

☐ E. 6

31) The following number line shows two points. Which value is located between these two numbers on the number line?

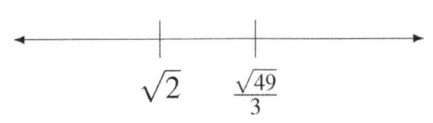

$\sqrt{2}$ $\frac{\sqrt{49}}{3}$

☐ A. 1

- B. 1.5
- C. 0
- D. 2.5
- E. 3

32) Which of the following graphs depicts y as a function of x?

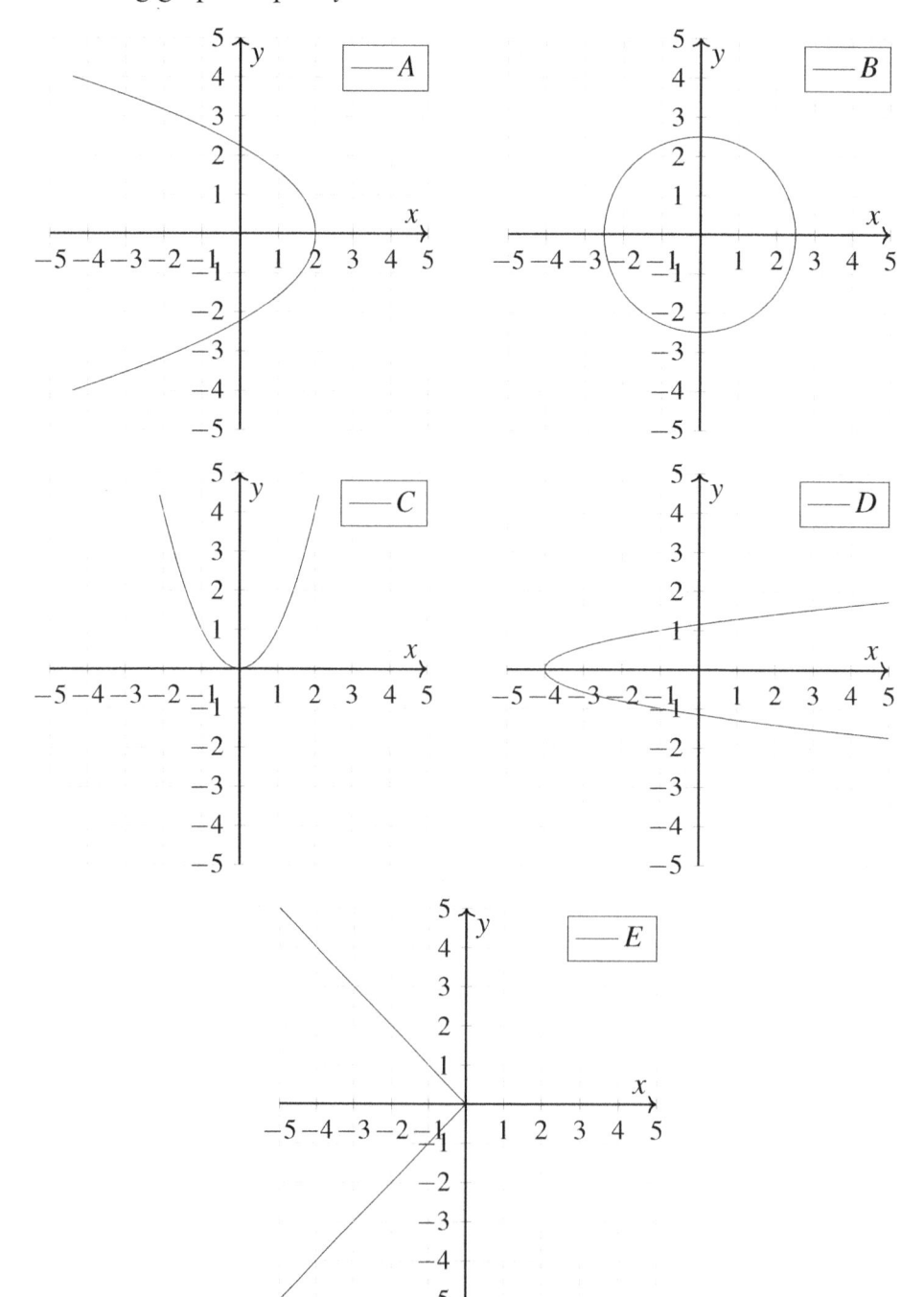

- A. Graph A

3.1 Practices

☐ B. Graph B

☐ C. Graph C

☐ D. Graph D

☐ D. Graph E

33) Determine the inequality statement that represents the set of possible values for the variable "v" satisfying the inequality $3v - 5y \leq 20$, where $y = 2$.

☐ A. $v \leq -10$

☐ B. $v \geq -10$

☐ C. $v \leq 10$

☐ D. $v \leq 20$

☐ E. $v \geq 20$

34) The graph of $y = -2x^2 + 8x + 4$ is depicted. If the graph intersects the y-axis at the point $(0, s)$, what is the value of s?

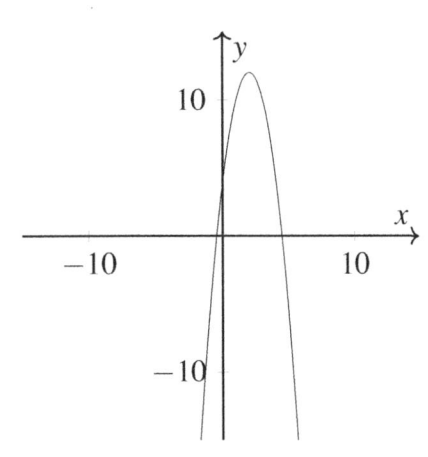

☐ A. -2

☐ B. 2

☐ C. 4

☐ D. 6

☐ E. 8

35) The mass of a tiny particle is 0.00005 kilograms. What is the scientific notation of this number?

☐ A. 5×10^{-4}

☐ B. 5×10^{-5}

☐ C. 5×10^4

☐ D. 5×10^5

☐ E. 5×10^6

36) Mark wants to buy a car and he needs a $4500 loan. Which loan option has the smallest amount of interest that he has to pay?

☐ A. A 15-month loan with a 6.00% annual simple interest rate

☐ B. A 20-month loan with a 5.75% annual simple interest rate

☐ C. A 24-month loan with a 5.50% annual simple interest rate

☐ D. A 30-month loan with a 6.25% annual simple interest rate

☐ E. A 36-month loan with a 5.25% annual simple interest rate

37) Consider the following two functions:

$$h(x) = 3x - 1,$$

$$j(x) = \frac{1}{2}x - 2.$$

How does the graph of h compare with the graph of j?

☐ A. The graph of h has the same y-intercept as the graph of j.

☐ B. The graph of h is parallel to the graph of j.

☐ C. The graph of h is less steep than the graph of j.

☐ D. The graph of h is steeper than the graph of j.

☐ E. The graph of h is perpendicular to the graph of j.

38) Kevin starts with $1200 in his investment account and adds $40 each week for y weeks. Linda starts with an investment account of $500 and adds $45 each week for y weeks.

Which inequality represents the situation when the amount of money in Kevin's account is greater than the amount of money in Linda's account?

☐ A. $40y < $45y - $1200

☐ B. $40y > $1200 + $45y

☐ C. $45y > $40y + $1200

☐ D. $45y < $700 + $40y

3.1 Practices

☐ E. $\$45y + \$40y < \$1200$

39) The cost to repair a car varies directly with the hours of labor required. Repairing a car for 3 hours costs $180. What is the cost, in dollars, for a 5-hour repair?

☐ A. $100

☐ B. $150

☐ C. $200

☐ D. $250

☐ E. $300

40) John has $10000 to invest in one of two accounts. Account C pays 5% simple interest per year, and account D pays 4.5% interest per year compounded annually. John will not make any more deposits or withdrawals. Which of these statements is true about the accounts after 4 years?

☐ A. Account C would earn John about $80.88 more interest than Account D.

☐ B. Account C would earn John about $74.81 more interest than Account D.

☐ C. Account D would earn John about $15.23 more interest than Account C.

☐ D. Account D would earn John about $20.12 more interest than Account C.

☐ E. Account C would earn John about $65 more interest than Account D.

41) Alice and Bob begin their careers at different firms. Alice earns $40000 a year at the start and gets a raise of $1800 every year. Bob earns $50000 a year at the start and gets a raise of $1300 every year. How many years will it take for Alice to catch up with Bob's salary?

☐ A. 7 years

☐ B. 10 years

☐ C. 12 years

☐ D. 15 years

☐ E. 20 years

42) A box contains numbered balls from 1 to 20. A ball is randomly selected. What is the probability that the ball picked is number 12?

☐ A. $\frac{1}{20}$

☐ B. $\frac{2}{20}$

- [ ] C. $\frac{10}{20}$
- [ ] D. $\frac{12}{20}$
- [ ] E. $\frac{19}{20}$

43) Consider the quadratic functions p and q given by:

$$p(x) = (x-d)^2 - 4,$$

$$q(x) = x^2 - 4x + 3.$$

Find the value of d such that the graph of q be 3 units above the graph of p?

- [ ] A. -4
- [ ] B. -3
- [ ] C. -2
- [ ] D. 2
- [ ] E. 4

44) In the coordinate plane, the plot of a certain polynomial function $g(x)$ crosses the x-axis exactly at two locations, denoted as $(c,0)$ and $(d,0)$. It is given that both c and d are negative values. Which of the following might be an appropriate representation for $g(x)$?

- [ ] A. $g(x) = (x-c)(x-d)$
- [ ] B. $g(x) = (x+c)(x+d)$
- [ ] C. $g(x) = (x-c)(x+d)$
- [ ] D. $g(x) = (x+c)(x-d)$
- [ ] E. $g(x) = x(x-c)(x-d)$

45) For the function $y = h(x)$, given that $h(-3) = 15$ and $h(9) = 8$, find the value of y when $x = 9$.

- [ ] A. $y = -15$
- [ ] B. $y = -9$
- [ ] C. $y = 8$
- [ ] D. $y = 9$
- [ ] E. $y = 15$

46) Among the given graphs, which one represents a line with an y-intercept of -1?

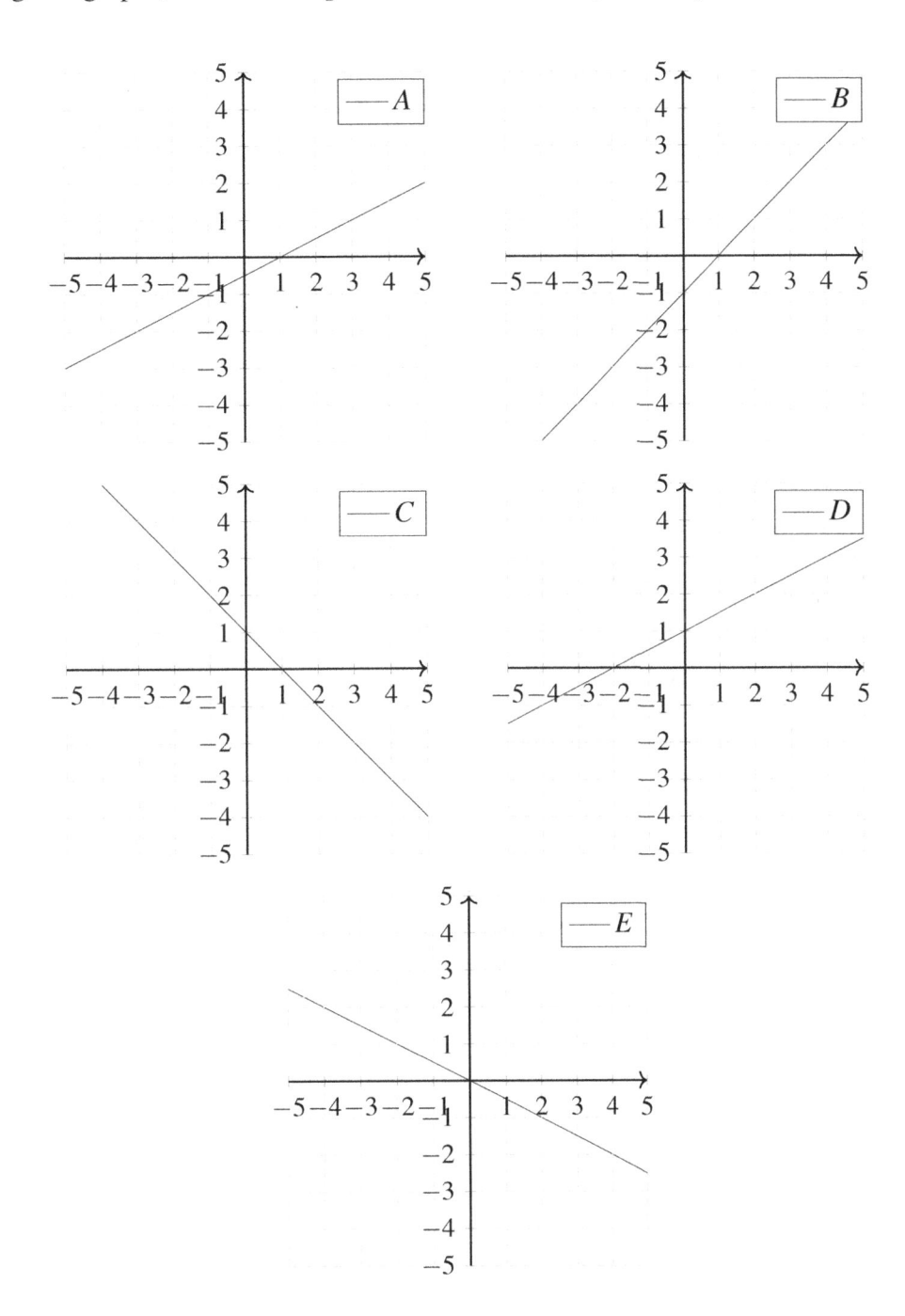

- A. Graph A
- B. Graph B
- C. Graph C
- D. Graph D
- D. Graph E

47) C. 2

48) D. 1300

49) C. 360

x, length $2x$, and height $3x$. What function represents the total area of all the external faces of these prisms combined?

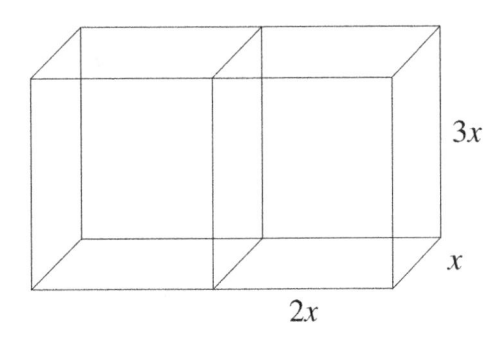

☐ A. $y = 20x^2$
☐ B. $y = 22x^2$
☐ C. $y = 28x^2$
☐ D. $y = 38x^2$
☐ E. $y = 24x^2$

3.2 Answer Keys

1) B. {(−1,0), (0,1), (1,2), (2,3)}
2) B. 3
3) D. 262
4) D. $14.78
5) C. $-\frac{4}{3}$ and $\frac{2}{3}$
6) D. School
7) D. $\sqrt{7}$
8) A. 25
9) A. Graph A
10) C. 5 cm, 12 cm, 13 cm
11) E. $11\frac{9}{20}$
12) B. 10
13) A. $5
14) B. (−1,−1)
15) B
16) C
17) E
18) A. 17
19) B. $5p = 10$
20) A
21) A
22) B. $y = 20x + 1100$
23) B
24) B. $600
25) B. $\frac{1}{2}n(n+5) + 6$

26) D. 3
27) A. $3052.08 cm^3$
28) B. $y = \frac{9}{5}x - \frac{2}{5}$
29) D. $\frac{20}{35}$
30) A. −6
31) B. 1.5
32) C. Graph C
33) C. $v \leq 10$
34) C. 4
35) B. 5×10^{-5}
36) A
37) D
38) D. $45y < \$700 + \$40y$
39) E. $300
40) B
41) E. 20 years
42) A. $\frac{1}{20}$
43) D. 2
44) A. $g(x) = (x-c)(x-d)$
45) C. $y = 8$
46) B. Graph B
47) C. 2
48) D. 1300
49) C. 360
50) D. $y = 38x^2$

3.3 Answers with Explanation

1) The set of ordered pairs that represents y as a function of x is set B. Each x-value in the set corresponds to exactly one y-value, and there are no repeated x-values.

2) The slope of a line that passes through the points (x_1, y_1) and (x_2, y_2) is calculated using the formula: $m = \frac{y_2 - y_1}{x_2 - x_1}$. Applying this formula to the given points $(2, -3)$ and $(7, 12)$, we have: $m = \frac{12 - (-3)}{7 - 2} = \frac{15}{5} = 3$. Thus, the slope of the line that passes through these points is 3, which is option B.

3) To find the total surface area (SA) of the given rectangular prism, we apply the formula:

$$SA = 2 \times (length \times width) + 2 \times (length \times height) + 2 \times (width \times height).$$

Plugging in the dimensions of the prism, we calculate:

$$SA = 2 \times (7 \times 5) + 2 \times (7 \times 8) + 2 \times (5 \times 8).$$

This simplifies to:
$$SA = 2 \times 35 + 2 \times 56 + 2 \times 40 = 262.$$

Therefore, the total surface area is 262 square inches.

4) For Account X with simple interest, the interest earned over 4 years is calculated by:

$$I_X = \text{Principal} \times \text{Rate} \times \text{Time} = \$1500 \times 0.04 \times 4 = \$240.$$

For Account Y with compound interest, the interest earned after 4 years is calculated by:

$$I_Y = \text{Principal} \times (1 + \text{Rate})^{\text{Time}} - \text{Principal} = \$1500 \times (1 + 0.04)^4 - \$1500 \approx \$254.78.$$

The difference in interest between Account X and Account Y is: $\$254.78 - \$240 = \$14.78$.

5) The zeros of the function are the x-values where the graph intersects the x-axis. For the function $g(x) = 6x^2 + 4x - 5$, the graph intersects the x-axis at two points, indicating two real roots. Since the graph crosses the x-axis on different sides of the y-axis, one zero is positive and the other is negative. Therefore, one of the two options A or C can be correct. But it is clear, from the graph, that the positive zero is smaller than 1 and the negative zero is smaller than -1. Therefore, option A cannot be correct either. The only option that matches the graph is option C.

6) Based on the coordinates given, the School appears to be located at $(3,-2)$ on the grid. This is deduced by understanding that the coordinate 3 indicates 3 units to the right from the origin, and -2 represents 2 units down from the origin.

7) Given the expression $x^2 + 3x + 2 = (x+2)(x+1)$, we can rewrite the fraction $x+2 = \frac{7(x+1)}{(x^2+3x+2)}$ as $x+2 = \frac{7(x+1)}{(x+2)(x+1)}$. Note that, since $x > -1$, the factor $x+1$ is not zero and we can remove it from the numerator and denominator. So, we have: $x+2 = \frac{7}{(x+2)}$. Simplifying, we find $(x+2)^2 = 7$. Thus, $x+2$ is either $\sqrt{7}$ or $-\sqrt{7}$ and D is the correct option.

8) Recognizing that a right angle measures $90°$, we set up the equation $4x - 10 = 90$ to find the value of x. Solving for x, we get $4x = 100$ and hence $x = 25$. So, the possible value for x is 25, which is option A.

9) The y-intercept is found when $x = 0$. Substituting $x = 0$ gives $y = -2$. So, the graph must cross the y-axis at $(0,-2)$. The x-intercept is found when $y = 0$. Setting $y = 0$ and solving for x gives $0 = \frac{1}{2}x - 2$, or $x = 4$. So, the graph must cross the x-axis at $(4,0)$. Graph A is the only graph that meets these criteria, making it the correct representation of the equation $y = \frac{1}{2}x - 2$.

10) The Pythagorean theorem states that for a right triangle, the sum of the squares of the two shorter sides equals the square of the longest side. Checking each option: For A, $2^2 + 3^2 = 4 + 9 = 13 \neq 6^2$. For B, $4^2 + 5^2 = 16 + 25 = 41 \neq 6^2$. For C, $5^2 + 12^2 = 25 + 144 = 169 = 13^2$, satisfying the theorem. For D, $7^2 + 8^2 = 49 + 64 = 113 \neq 9^2$. For E, $8^2 + 9^2 = 64 + 81 = 145 \neq 11^2$. Hence, the correct option is C.

11) First, convert all fractions to a common denominator (20):

$$2\frac{12}{20} + 1\frac{5}{20} + 4\frac{2}{20} + 3\frac{10}{20} = (2+1+4+3) + \left(\frac{12+5+2+10}{20}\right) = 10 + \frac{29}{20} = 10 + 1\frac{9}{20} = 11\frac{9}{20}.$$

12) Let the number of movies watched be x. For service A we have: $10 + 0.50x$. For service B we get: $5 + 1x$. Equating the two costs: $10 + 0.50x = 5 + 1x$. Solving for x: $5 = 0.50x$, hence $x = 10$. After watching 10

3.3 Answers with Explanation

movies, the costs for both services will be the same, which is option B.

13) Let the cost of each ticket be t. For Class A: $10t - 50$. For Class B: $5t - 25$. Since both spent the same amount: $10t - 50 = 5t - 25$. Solving for t: $5t = 25$, hence $t = 5$. The cost of each ticket was $5, so the option A is correct.

14) Two equations will have a common solution if their graphical representations intersect at a single point on the coordinate grid. As per the given diagram, the intersection point of the two lines is at $(-1, -1)$. Consequently, option B is identified as the correct answer.

15) A rotation of $180°$ about the origin is a rigid transformation that maintains the shape and size of the figure. Therefore, $DEFG$ is congruent to $D'E'F'G'$, and their angles and side lengths are equal. The sum of the angles, the area, and the perimeter remain unchanged. Therefore, statement B is not true as the angle measures of $DEFG$ are not greater than those of $D'E'F'G'$.

16) For Tom, who already swam 30 laps, the equation representing the total number of laps swum after x hours is $4x + 30$. For Jerry, swimming 8 laps per hour, the equation is $8x$. Setting these equal: $4x + 30 = 8x$, which matches the given equation. Therefore, option C correctly represents the equation. The equations for A, B, D and E are $4x = 8x$, $8x + 30 = 4x$, $8x = 30x$ and $30x = 8x$, respectively.

17) To find the equation of $h(x)$, we use two points $(0, 3)$ and $(3, 6)$. The slope m is $\frac{6-3}{3-0} = 1$. Thus, $h(x) = x + 3$. Translating h right by 4 units to create j gives $j(x) = h(x - 4)$. So, $j(x) = (x - 4) + 3 = x - 1$. The y-intercept of h is 3 and of j is -1. Therefore, the y-intercept of h is 4 units above of j's y-intercept, making E correct.

18) The slide, being the hypotenuse of the right triangle, can be calculated using the Pythagorean theorem: $a^2 = b^2 + c^2$, where a is the length of the hypotenuse, and b and c are the lengths of the other two sides. Given $b = 8$ meters and $c = 15$ meters, we have: $a^2 = 8^2 + 15^2 \Rightarrow a^2 = 64 + 225 \Rightarrow a^2 = 289 \Rightarrow a = 17$. Therefore, the length of the slide is 17 meters, which corresponds to option A.

19) Let p represents the cost of each pencil. Since the total cost for 5 pencils is $10, the equation is: $5p = 10$.

20) A. Only the graph of $g(x)$ has a maximum point. The graphs of $f(x)$ and $h(x)$ have minimum points.

B. The axis of symmetry for all three functions is the vertical line $x = 0$.

C. The graphs of $g(x)$ and $h(x)$ cross the x-axis.

D. The y-intercepts of the functions $f(x)$, $g(x)$ and $h(x)$ are 4, 5, and -6, respectively, and are therefore different.

E. As can be seen in the graph, the functions $g(x)$ and $h(x)$ intersect at two points.

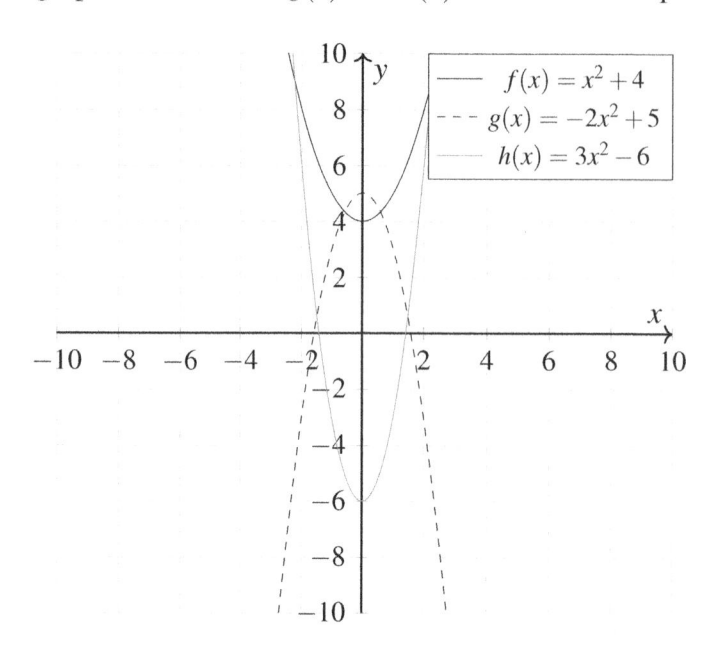

21) To determine the transformation from the graph of $h(x)$ to $k(x)$, we will visually analyze the changes in slope and y-intercept based on the given graph.

- Change in Slope: Observe the steepness of the lines. The line representing $h(x)$ slopes downward, indicating a negative slope. In contrast, the line for $k(x)$ slopes upward more steeply, indicating a positive slope that is steeper than the negative slope of $h(x)$. This suggests that the slope of $k(x)$ is a negative multiple of the slope of $h(x)$, which is consistent with multiplying the slope by -2.

- Change in y-intercept: The y-intercept is where each line crosses the y-axis. The line for $h(x)$ crosses the y-axis above the origin, while the line for $k(x)$ crosses below the origin. This indicates that the y-intercept for $k(x)$ is lower than that for $h(x)$, suggesting a decrease in the y-intercept value. The magnitude of this change is consistent with a decrease by 8 units.

Therefore, based on the graph, the transformation from $h(x)$ to $k(x)$ involved multiplying the slope by -2 and decreasing the y-intercept by 8 units. Option A is the correct choice reflecting these changes.

22) To find the function, we use the given data points $(35, 1800)$ and $(70, 2500)$ to calculate the slope: $m = \frac{2500-1800}{70-35} = \frac{700}{35} = 20$. Using the slope and one point, say $(35, 1800)$, to find the y-intercept through $y = mx + b$: $1800 = 20 \times 35 + b$, solving for b gives $b = 1100$. The function is $y = 20x + 1100$, which is the

3.3 Answers with Explanation

option B.

23) Scatterplot B shows a trend where stress levels decrease as weekly physical activity increases, which supports the study's findings. The negative correlation in Scatterplot B, where higher values of weekly physical activity are associated with lower stress levels, aligns with the study's suggestion. Scatterplots A, C, D and E do not show this trend and therefore do not support the findings.

24) Applying the formula for simple interest $I = Prt$, where I represents the interest, P the principal amount, r the rate of interest, and t the time in years, the calculation follows: $I = \$3000 \times 0.04 \times 5 = \$600$. Thus, the accrued interest over a period of 5 years amounts to $600.

25) The sequence shows that each term is greater than the previous one by an incrementally increasing amount (4, 5, 6, 7, etc.). This pattern indicates a quadratic nature. The correct formula can be deduced by testing the given options. For $n = 1$, $\frac{1}{2}(1)(1+5)+6 = 9$; for $n = 2$, $\frac{1}{2}(2)(2+5)+6 = 13$; continuing this pattern matches the given sequence.

26) By analyzing the graph data, we note that the function g dips to its lowest point at the coordinates $(3, -4)$. This indicates that the function $g(x)$ attains its minimum value at $x = 3$. Consequently, the correct answer is option D.

27) To determine the volume of the ball, we use the sphere volume formula $V = \frac{4}{3}\pi r^3$, where r is half of the diameter. Thus, with a diameter of 18cm, the radius is 9cm. Inserting the radius into the formula yields: $V = \frac{4}{3} \times 3.14 \times 9^3 = 3052.08 cm^3$.

28) To find the equation of a line passing through two points (x_1, y_1) and (x_2, y_2), we use the formula $y = mx + b$. First, calculate the slope (m) using the slope formula: $m = \frac{y_2 - y_1}{x_2 - x_1}$. Substituting the given points $(3, 5)$ and $(-2, -4)$: $m = \frac{-4-5}{-2-3} = \frac{-9}{-5} = \frac{9}{5}$. The equation becomes: $y = \frac{9}{5}x + b$. To find b, substitute one of the points, say $(3, 5)$: $5 = \frac{9}{5} \times 3 + b$. Solving for b, we get: $b = 5 - \frac{27}{5} = -\frac{2}{5}$. So, the equation is: $y = \frac{9}{5}x - \frac{2}{5}$.

29) From the table, we observe that in Group 2, 20 students are working with paint. The total number of students in Group 2 is 35. Hence, the ratio of students in Group 2 who are painting, as compared to the total in that group, is represented by $\frac{20}{35}$. This fraction reflects the portion of Group 2 engaged in painting activities. Option D is the accurate representation of this ratio.

30) To solve the equation $5x - 8 = 7x + 4$, first rearrange the terms to isolate x: $5x - 8 = 7x + 4 \Rightarrow 5x - 7x = 4 + 8 \Rightarrow -2x = 12$. Now, solve for x: $x = -6$.

31) We need to compare the values of $\sqrt{2}$ and $\frac{\sqrt{49}}{3}$ on the number line. $\sqrt{2}$ is approximately 1.41 and $\frac{\sqrt{49}}{3} = \frac{7}{3} \approx 2.33$. Looking at the options, the number that falls between 1.41 and 2.33 is 1.5. Therefore, the correct answer is option B.

32) A graph represents y as a function of x if it passes the vertical line test. This means that any vertical line drawn on the graph should intersect it at no more than one point. Options A, B, D, and E fail the vertical line test at certain points. The only graph that consistently passes the vertical line test is a parabola opening upwards (option C), as it intersects any vertical line at exactly one point. Thus, the correct answer is option C.

33) Substituting $y = 2$ into the inequality $3v - 5y \leq 20$ gives $3v - 5(2) \leq 20$. Simplifying this, we get $3v - 10 \leq 20$. Adding 10 to both sides yields $3v \leq 30$. Dividing both sides by 3, we find $v \leq 10$, which is option C.

34) Since the graph intersects the y-axis at $(0, s)$, we substitute 0 for x and s for y in the equation $y = -2x^2 + 8x + 4$. This results in: $s = -2(0)^2 + 8(0) + 4$, which simplifies to $s = 4$. Therefore, the correct answer is option C, where s is 4.

35) To convert 0.00005 to scientific notation, we move the decimal point five places to the right. This gives us 5 as the coefficient. Since we moved the decimal five places, we multiply by 10 to the power of -5, representing the number of decimal places moved. Therefore, the correct representation in scientific notation is 5×10^{-5}, making the correct answer option B.

36) To find the loan option with the least interest, we use the simple interest formula: $I = P \times r \times t$, where I is the interest, P is the principal (loan amount), r is the annual interest rate in decimal form, and t is the time in years. Calculating the interest for each option:

Option A: $I = 4500 \times 0.06 \times \frac{15}{12} = \$337.50$.

Option B: $I = 4500 \times 0.0575 \times \frac{20}{12} = \$431.25$.

Option C: $I = 500 \times 0.055 \times 2 = \$495$.

Option D: $I = 4500 \times 0.0625 \times \frac{30}{12} = \$703.13$.

Option E: $I = 4500 \times 0.0525 \times 3 = \$708.75$.

3.3 Answers with Explanation

The lowest interest amount is with Option A, making it the correct choice.

37) To compare the graphs of $h(x) = 3x - 1$ and $j(x) = \frac{1}{2}x - 2$, we consider their slopes and y-intercepts. The slope of $h(x)$ is 3. The slope of $j(x)$ is $\frac{1}{2}$. Since the slope of $h(x)$ is greater in magnitude than that of $j(x)$, the graph of h is steeper than the graph of j. Therefore, the correct answer is D.

The y-intercepts are also different: $h(x)$ intersects the y-axis at $(0, -1)$ and $j(x)$ at $(0, -2)$, confirming that answer A is incorrect. As the slopes of $h(x)$ and $j(x)$ differ, the graphs are not parallel (ruling out B), and they are not perpendicular either (ruling out E).

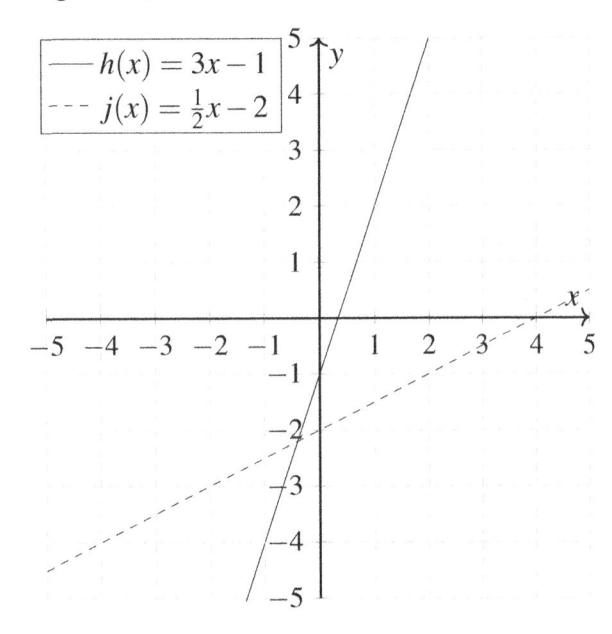

38) To find the total in Kevin's account: $\$1200 + \$40y$. For Linda: $\$500 + \$45y$. To compare: $\$1200 + \$40y > \$500 + \$45y$. Simplifying gives: $\$45y < \$700 + \$40y$. Thus, D is correct.

39) Let y be the cost and x the hours. Given $y = kx$ with $180 = k \cdot 3$, we find $k = 60$. For a 5-hour repair: $y = 60 \times 5 = \$300$. Thus, E is correct.

40) For Account C with simple interest, the interest earned over 4 years is calculated by:

$$I_C = \text{Principal} \times \text{Rate} \times \text{Time} = 10000 \times 0.05 \times 4 = 2000.$$

For Account D with compound interest, the interest earned after 4 years is calculated by:

$$I_D = \text{Principal} \times (1 + \text{Rate})^{\text{Time}} - \text{Principal} = 10000 \times (1 + 0.045)^4 - 10000 \approx 1925.19.$$

The difference in interest between Account C and Account D is: $2000 - $1925.19 \approx 74.81. Thus, B is correct.

41) Let y be the number of years. Alice's salary: $40000 + 1800y$, Bob's salary: $50000 + 1300y$. Set them equal: $40000 + 1800y = 50000 + 1300y$. Solving gives $y = 20$. Thus, E is correct.

42) The furmula of the Probability is: $\frac{\text{number of desired outcomes}}{\text{number of total outcomes}}$. Total balls (number of total outcomes): 20. Balls with number 12 (number of desired outcomes): 1. Probability of picking 12: $\frac{1}{20}$. Thus, A is correct.

43) Rewriting $q(x)$: $q(x) = x^2 - 4x + 3 = (x^2 - 4x + 4) - 1 = (x-2)^2 - 1$. Since $p(x)$ is 3 units below $q(x)$, set $p(x) = q(x) - 3 \Rightarrow p(x) = ((x-2)^2 - 1) - 3 = (x-2)^2 - 4$. Comparing with $p(x) = (x-d)^2 - 4$, we find $d = 2$. Thus, D is correct.

44) For $g(x)$ to cross at $(c, 0)$ and $(d, 0)$, it must be true that $g(c) = 0$ and $g(d) = 0$. $g(x) = (x-c)(x-d)$ satisfies this, as $g(c) = (c-c)(c-d) = 0$ and $g(d) = (d-c)(d-d) = 0$. Thus, A is correct.

45) Substituting $x = 9$ into $h(x)$ gives $y = h(9)$. Given $h(9) = 8$, thus $y = 8$. Therefore, C is correct.

46) The y-intercept of a graph is the point where it crosses the y-axis. The correct graph is the one that intersects the y-axis at the point $(0, -1)$. Among the provided options, the graph corresponding to option B meets this criterion.

47) To solve, we need to evaluate the expression using the given values of $k(x)$. According to the table, $k(-1) = 2$ and $k(2) = -1$. Substitute these into the expression: $3 - (-1) + 2(-1) = 3 + 1 - 2$. Simplifying gives: $4 - 2 = 2$. Thus, the correct answer is option C.

48) The task is to determine the total number of packages transported after a certain number of conveyor runs. The relationship between the number of runs and packages is linear, as each run carries the same number of packages. First, establish the rate of package transport per run. Using the provided data, with 10 runs carrying 1000 packages, the rate per run is calculated as $1000 \div 10 = 100$ packages per run. The general formula for the total number of packages y after x runs is $y = 100x$, where 100 is the number of packages per run. To find the total for 13 runs, substitute $x = 13$ into the formula: $y = 100 \times 13 = 1300$. Thus, the conveyor transports 1300 packages in 13 runs, corresponding to option D.

49) Since $x = 10$ is a solution, substitute 10 for x in the equation: $(4 \times 10 + d)^2 = 160000$. This simplifies to

3.3 Answers with Explanation

$(40+d)^2 = 160000$. Taking the square root of each side gives two equations: $40+d = 400$ and $40+d = -400$. Solving the first equation for d yields $d = 360$, and the second equation yields $d = -440$. Thus, one possible value of d is 360.

50) The surface area of one rectangular prism with dimensions x, $2x$, and $3x$ is calculated by summing the area of each face. The prism has 6 faces, with areas $2x \times 3x$, $x \times 3x$, and $x \times 2x$, each counted twice. Thus, the total area for one prism is $2(6x^2 + 3x^2 + 2x^2) = 22x^2$. With two such prisms attached, one face from each prism will be internal and not counted. This removes two faces of area $x \times 3x = 3x^2$ in total. So, the total external area for both prisms is $(4 \times 6x^2) + (2 \times 3x^2) + (4 \times 2x^2) = 38x^2$.

4. Practice Test 3

CBEST Math Practice Test

Total number of questions: 50

Total time: 90 Minutes

Calculators are prohibited for the CBEST exam.

4.1 Practices

1) The vertices of a rectangular are located at $P = (5,3)$, $Q = (5,-4)$, $R = (-3,-4)$, and $S = (-3,3)$. Choose the closest value to the distance between the points Q and S.

4.1 Practices

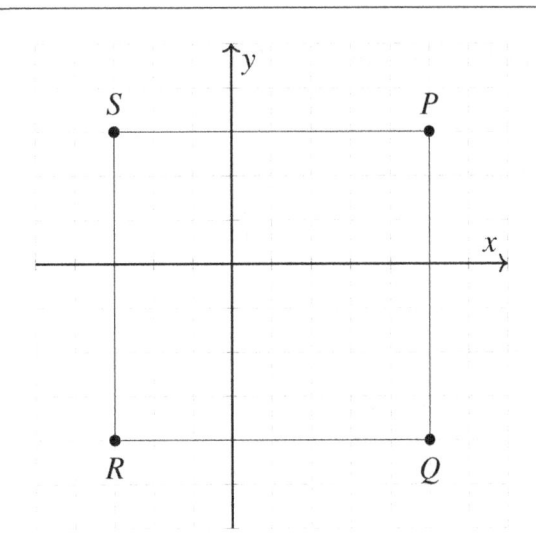

[Figure: Rectangle with vertices S (upper left), P (upper right), R (lower left), Q (lower right) on a coordinate plane]

- [ ] A. 9 units
- [ ] B. 9.5 units
- [ ] C. 12 units
- [ ] D. 11 units
- [ ] E. 15 units

2) A line passes through $(4,3)$ and $(1,-2)$. Determine the y-intercept of this line.

- [ ] A. -1
- [ ] B. $-\frac{11}{3}$
- [ ] C. 1
- [ ] D. $\frac{11}{3}$
- [ ] E. $\frac{4}{3}$

3) A cylindrical water tank measures 18 meters in height and has a radius of 3 meters. What is the total volume of the tank in cubic meters? Consider $\pi = 3.14$.

- [ ] A. 152.7 m^3
- [ ] B. 305.4 m^3
- [ ] C. 508.68 m^3
- [ ] D. 611.5 m^3
- [ ] E. 1223 m^3

4) Jamie plans to deposit $3000 in a bank offering 3% simple annual interest (Account A) or another bank

offering 2.8% compounded annually (Account B). After 5 years, which account will have accrued more interest?

☐ A. Account A would earn Jamie about $6 more than Account B.

☐ B. Account B would earn Jamie about $10 more than Account A.

☐ C. Account A would earn Jamie about $90 more than Account B.

☐ D. Account B would earn Jamie about $20 more than Account A.

☐ E. Both accounts would earn the same amount.

5) Identify the equation that describes a direct proportional relationship:

☐ A. $y = \frac{1}{2}x$

☐ B. $y = 4$

☐ C. $y = -2x^2 + 3$

☐ D. $y = 3x^2$

☐ E. $y = \frac{1}{x}$

6) Five years ago, Julia placed a sum of $2000 into a savings account. This account accrues interest at a simple annual rate of 5%. With no further deposits or withdrawals, calculate the total amount in Julia's account at the end of this five-year period.

☐ A. $500

☐ B. $750

☐ C. $2500

☐ D. $2750

☐ E. $3000

7) Consider a circle positioned on a coordinate plane, with its center located at the coordinates $(3, -4)$. The circle is subjected to a translation of m units to the right and n units downward. Determine the transformation rule that describes the new position of the circle's center following this translation.

☐ A. $(x, y) \to (3 + m, -4 - n)$

☐ B. $(x, y) \to (3 - m, -4 + n)$

☐ C. $(x, y) \to (-3 + m, -4 - n)$

☐ D. $(x, y) \to (-3 - m, -4 + n)$

☐ E. $(x, y) \to (3 + m, -4 + n)$

4.1 Practices

8) In a class of 50 students, 40% are proficient in a foreign language. How many students are not proficient in that language?

- ☐ A. 15
- ☐ B. 20
- ☐ C. 30
- ☐ D. 35
- ☐ E. 40

9) In a school, each student takes different courses each with grades ranging from 0 to 100. Mark's average grade in 6 courses is 70. Emily took 7 courses. If the total sum of their course grades is the same, what is Emily's average course grade?

- ☐ A. 46.67
- ☐ B. 52.22
- ☐ C. 60
- ☐ D. 70
- ☐ E. 78.33

10) Two school clubs, Club Alpha and Club Beta, are redistributing their members. Club Alpha has 90 members, and Club Beta has 40 members. Club Alpha will transfer a number of members to Club Beta in multiples of 5, until both clubs have the same number of members. How many members must Club Alpha transfer to Club Beta?

- ☐ A. 5
- ☐ B. 15
- ☐ C. 20
- ☐ D. 25
- ☐ E. 30

11) Michael is painting a fence and completes 8 feet in the first 4 days. If Michael continues painting at this rate, which graph represents the length of the fence painted in feet per day?

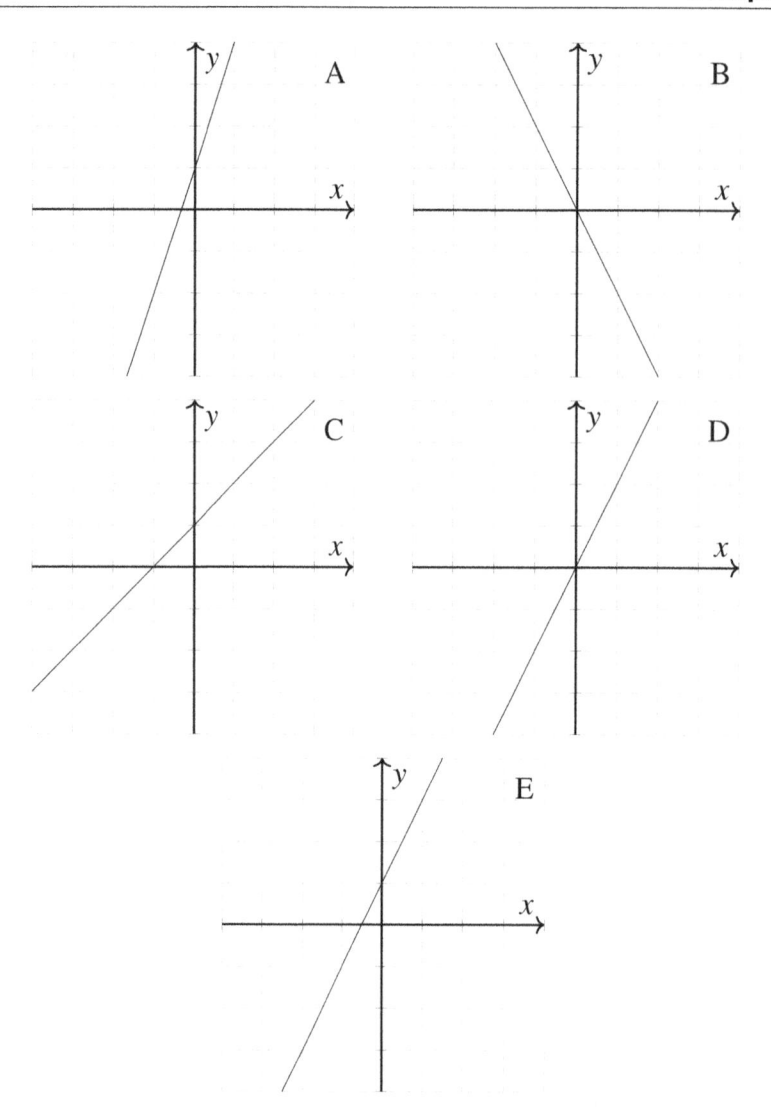

- ☐ A. Graph A
- ☐ B. Graph B
- ☐ C. Graph C
- ☐ D. Graph D
- ☐ E. Graph E

12) On a coordinate grid, a shape undergoes a transformation to form a new figure. Which of the following transformations will alter the size but not the orientation of the original figure?

- ☐ A. Rotating the figure by 180° counterclockwise
- ☐ B. Reflecting the figure across the y-axis
- ☐ C. Scaling the figure by a factor of 0.5
- ☐ D. Reflecting the figure across the line $y = x$

4.1 Practices

☐ E. Shifting the figure 5 units upwards

13) A graph shows the relationship between the cost of cooking classes and the duration of the classes at two culinary schools.

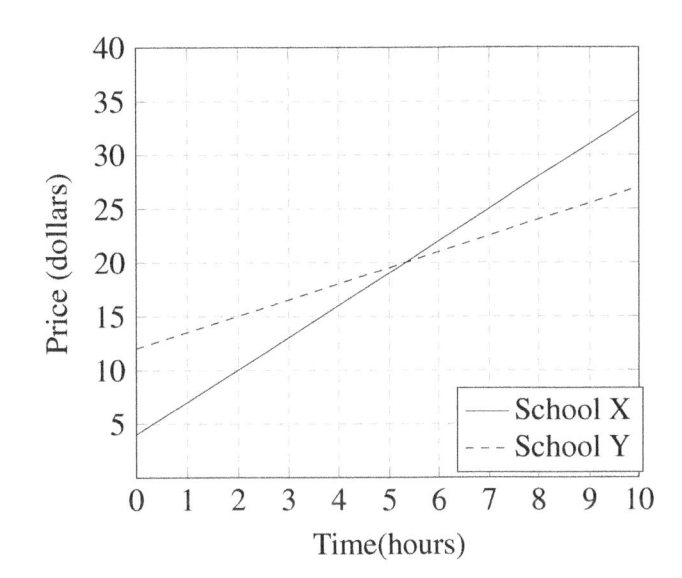

Which statement about the cost of a 9-hour class is supported by the graph?

☐ A. A 9-hour class at School X is about $15 more expensive than at School Y.

☐ B. A 9-hour class at School X is about $15 less expensive than at School Y.

☐ C. A 9-hour class at School X is about $5 more expensive than at School Y.

☐ D. A 9-hour class costs the same at both School X and School Y.

☐ E. It is impossible to determine the cost of a 9-hour class for either School X or School Y from the graph.

14) Which set of the following ordered pairs correctly defines a function?

☐ A. $\{(1,2), (2,3), (3,5), (1,4)\}$

☐ B. $\{(4,6), (8,2), (4,7), (6,3)\}$

☐ C. $\{(2,4), (3,6), (4,8), (5,10)\}$

☐ D. $\{(7,3), (8,3), (8,4), (-2,4)\}$

☐ E. $\{(3,6), (5,10), (7,14), (3,18)\}$

15) What is the value of y-intercept, for the linear function $y = 3x + 2$?

☐ A. -3

☐ B. -2

- [ ] C. 2
- [ ] D. 3
- [ ] E. 5

16) A cylindrical can has a diameter of 4 inches and a height of 6 inches. Which formula calculates V, the volume of the can in cubic inches?

- [ ] A. $V = \pi(2)^2(3)$
- [ ] B. $V = \pi(2)^2(6)$
- [ ] C. $V = \pi(4)^2(3)$
- [ ] D. $V = \pi(4)^2(6)$
- [ ] E. $V = \pi(3)^2(4)$

17) A square on a coordinate grid will be shrunk to create a smaller square, using the origin as the center of shrinkage. Which rule represents this transformation?

- [ ] A. $(x,y) \to (1.75x, 1.75y)$
- [ ] B. $(x,y) \to (x+5, y+5)$
- [ ] C. $(x,y) \to (1.2x, 1.2y)$
- [ ] D. $(x,y) \to (0.8x, 0.8y)$
- [ ] E. $(x,y) \to (x-0.15, y-0.15)$

18) A ribbon that is $3\frac{4}{5}$ feet long is cut into two segments. The longer piece measures y feet. Which inequality represents all possible values of y?

- [ ] A. $y > 1\frac{9}{10}$
- [ ] B. $y > 2\frac{2}{5}$
- [ ] C. $y < 3$
- [ ] D. $y < 3\frac{2}{5}$
- [ ] E. $y < 3\frac{4}{5}$

19) Tom has a bag containing 4 orange balls, 6 blue balls, and 2 green balls. If he picks one ball at random, what is the probability it will not be orange?

- [ ] A. $\frac{4}{12}$
- [ ] B. $\frac{6}{12}$

4.1 Practices

☐ C. $\frac{8}{12}$

☐ D. $\frac{10}{12}$

☐ E. $\frac{11}{12}$

20) George chooses to invest $1200 into two savings accounts. Account X earns 2.5% annual simple interest. Account Y earns 2.5% interest, compounded annually. George makes no additional deposits or withdrawals. which option shows the nearest valuu to the sum of the balances in Account X and Account Y at the end of 3 years?

☐ A. $1290

☐ B. $1389.19

☐ C. $1138.4

☐ D. $2582.27

☐ E. $2400

21) Mark is designing a rectangular poster that measures 20 cm in length and 15 cm in width. Determine the length of the diagonal of this poster in centimeters.

☐ A. 25 cm

☐ B. 30 cm

☐ C. 35 cm

☐ D. 40 cm

☐ E. 625 cm

22) In a triangle, the ratio of the angles is 2 : 3 : 4. Determine the measure of the largest angle in the triangle.

☐ A. 36°

☐ B. 48°

☐ C. 72°

☐ D. 80°

☐ E. 108°

23) A boulder has a mass of 2500 kilograms. Express this mass in scientific notation.

☐ A. 2.5×10^{-3}

☐ B. 2.5×10^2

☐ C. 2.5×10^3

☐ D. 2.5×10^4

☐ E. 2.5×10^5

24) What is the domain of the portion of the exponential function displayed on the coordinate grid?

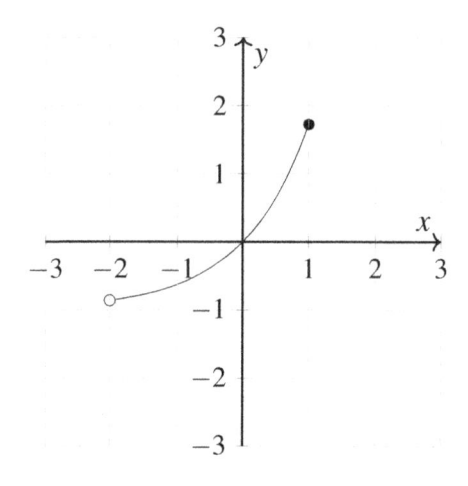

☐ A. $-1 < x \leq 2$

☐ B. $-1 \leq x \leq 2$

☐ C. $-2 < x \leq 1$

☐ D. $-2 \leq x \leq 1$

☐ E. $-0.5 < x \leq 2$

25) Among the given representations, which one shows y as a function of x?

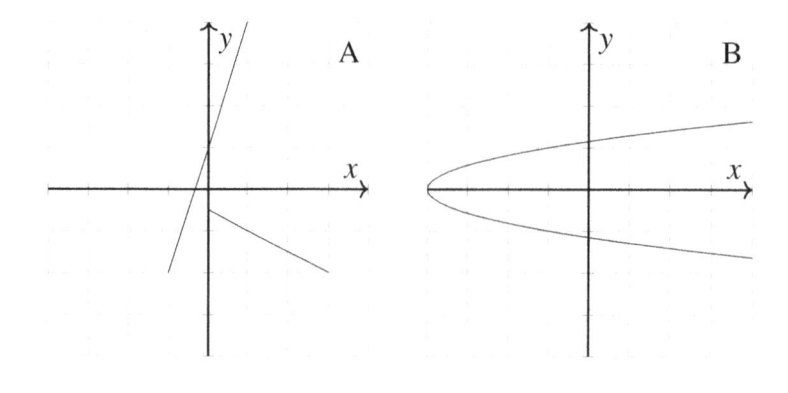

4.1 Practices

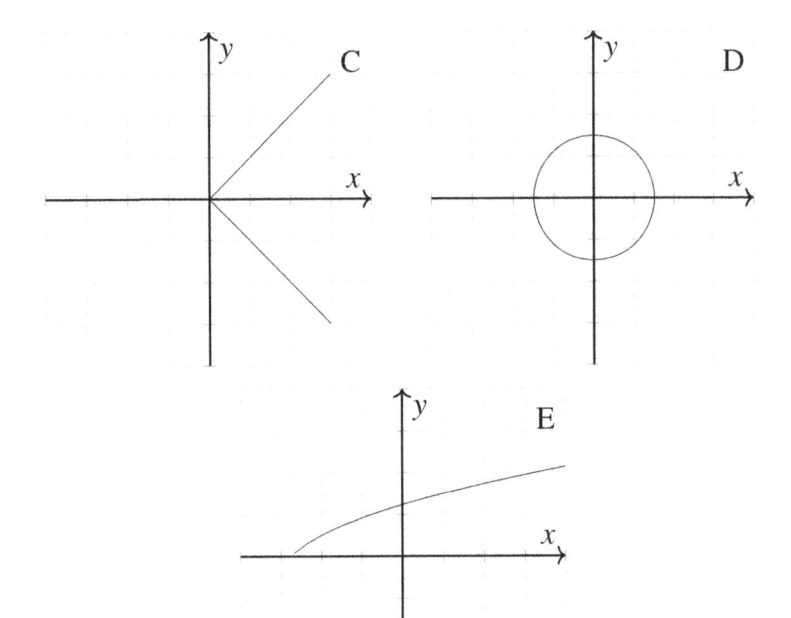

☐ A. Graph A
☐ B. Graph B
☐ C. Graph C
☐ D. Graph D
☐ E. Graph E

26) Linda bought a book for $18. This cost was $6 less than twice the amount she spent on a notebook. How much did the notebook cost?

☐ A. $9
☐ B. $10
☐ C. $12
☐ D. $14
☐ E. $15

27) The table below shows the linear relationship between the distance traveled by a taxi in miles and the fare charged in dollars.

Distance traveled(x)	Fare charged(y)
2	10
4	16
6	22
8	28
10	34

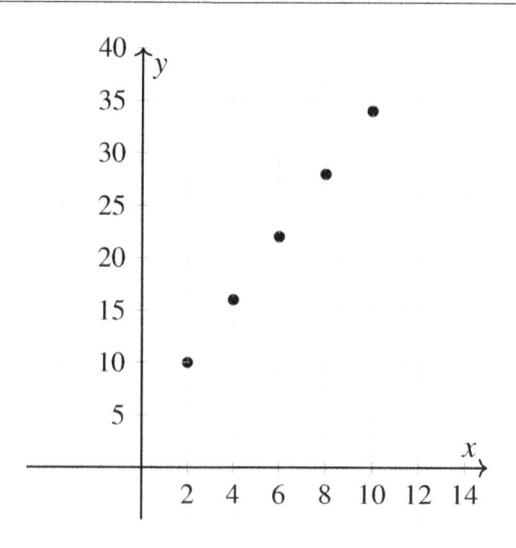

What is the slope of the line that represents this relationship?

☐ A. $2 per mile

☐ B. $3 per mile

☐ C. $4 per mile

☐ D. $5 per mile

☐ E. $6 per mile

28) In the following figure, what is the value of *y* in the given triangle?

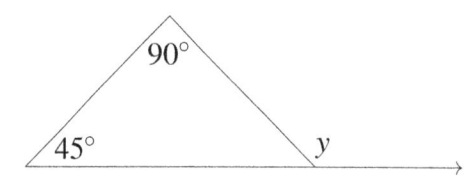

☐ A. $y = 45°$

☐ B. $y = 55°$

☐ C. $y = 90°$

☐ D. $y = 135°$

☐ E. $y = 225°$

29) If $b = 2$ and $\frac{a}{4} = b$, what is the value of $a^2 + 4b$?

☐ A. 66

☐ B. 70

☐ C. 72

☐ D. 76

4.1 Practices

☐ E. 81

30) Identify which point among the following options is located on the line defined by the equation $5x+2y=14$.

☐ A. (0,7)

☐ B. (2,3)

☐ C. (1,4)

☐ D. (3,2)

☐ E. (4,1)

31) A plant nursery sells decorative pots for w dollars each. Which scenario aligns best with the inequality $7w < 5w+6$?

☐ A. The price for 7 pots is more than the cost of 5 pots and a $6 plant.

☐ B. Acquiring 7 pots is less expensive than 5 pots with a $6 reduction.

☐ C. The expense for 7 pots is lower than purchasing 5 pots and a $6 plant.

☐ D. The total for 7 pots and a $6 plant is under the price of 5 pots.

☐ E. The amount for 7 pots surpasses the cost of 5 pots by $6.

32) In the diagram provided, a certain region is shaded. This region is bounded by the graphs of $x \geq 0$, $y \geq 0$, and which other inequality?

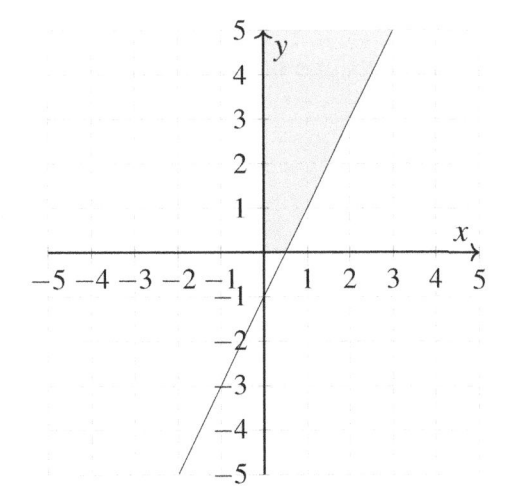

☐ A. $y \geq x+1$

☐ B. $y \leq 2x+1$

☐ C. $y \geq x-1$

- D. $y \geq 2x - 1$
  - E. $y \leq x - 1$

33) Jessica starts with $300 in her account and deposits $25 each week for *y* weeks. Mark starts with no money in his account but deposits $40 each week for *y* weeks. Which inequality represents the situation when the amount in Jessica's account is more than the amount in Mark's account?

  - A. $25y < 300 + 40y$
  - B. $25y > 300 + 40y$
  - C. $40y < 300 + 25y$
  - D. $40y > 300 + 25y$
  - E. $40y + 25y < 300$

34) Given $d = 3$ and $\frac{c}{5} = d$, calculate the value of $c + 5d^2$.

  - A. 54
  - B. 58
  - C. 60
  - D. 69
  - E. 70

35) Triangles *JKL* and *MNO* are similar right triangles placed on a coordinate plane. Identify the proportion that shows the slope of $\overline{JK}$ is the same as the slope of $\overline{MN}$.

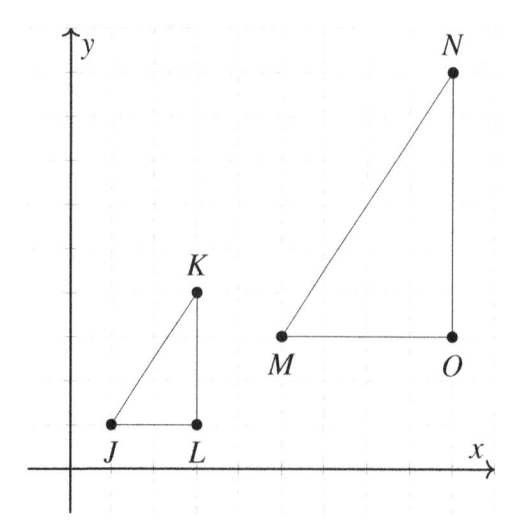

  - A. $\frac{\overline{JK}}{\overline{JL}} = \frac{\overline{MN}}{\overline{MO}}$

4.1 Practices

- [ ] B. $\frac{JL}{JK} = \frac{MO}{MN}$
- [ ] C. $\frac{KL}{JL} = \frac{NO}{MO}$
- [ ] D. $\frac{KL}{JK} = \frac{NO}{MN}$
- [ ] E. $\frac{JL}{KL} = \frac{MO}{NM}$

36) A club is purchasing t-shirts for its members. There is a setup fee of $15 and each t-shirt costs $4. What function shows the total cost in dollars for buying n t-shirts?

- [ ] A. $T = 4n + 15$
- [ ] B. $T = 4n - 15$
- [ ] C. $T = 15n - 4$
- [ ] D. $T = 15n + 4$
- [ ] E. $T = 15n$

37) Calculate the total surface area of a cuboid-shaped container with dimensions 7 meters in length, 5 meters in width, and 3 meters in height.

- [ ] A. 130 m^2
- [ ] B. 142 m^2
- [ ] C. 150 m^2
- [ ] D. 166 m^2
- [ ] E. 174 m^2

38) A spherical balloon has a diameter of 6 feet. What is the volume of air it can contain when fully inflated, in cubic feet? Consider $\pi = 3.14$.

- [ ] A. 113.10 ft^3
- [ ] B. 150.08 ft^3
- [ ] C. 200.96 ft^3
- [ ] D. 113.04 ft^3
- [ ] E. 339.29 ft^3

39) Consider the test scores of a student over a series of exams: 82, 76, 90, 85, 78, 90, 82, 88, 84, 79. What is the mean absolute deviation (MAD) of these scores?

- [ ] A. 2.4

☐ B. 3.6

☐ C. 4

☐ D. 5.4

☐ E. 6.2

40) The graph illustrates the linear relationship between the distance covered (in meters) by a drone and the amount of battery power used (in hours). Determine the best representation of the rate of change of the distance covered with respect to the amount of battery power used.

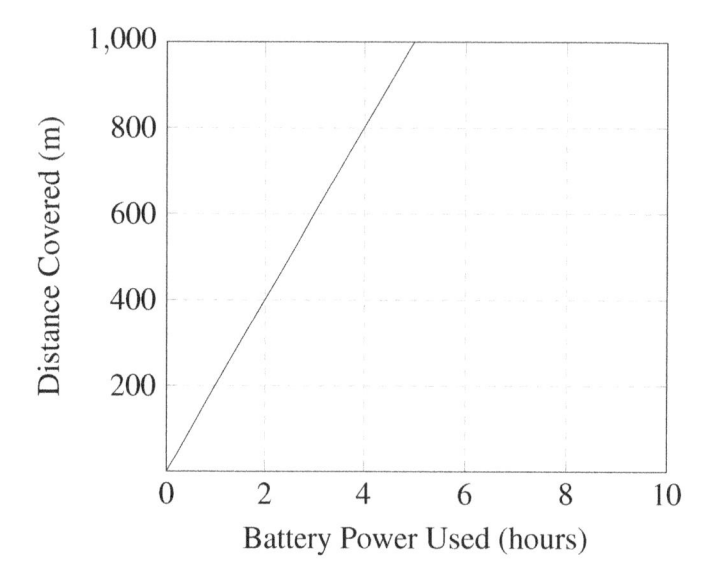

☐ A. $50 \frac{m}{hr}$

☐ B. $100 \frac{m}{hr}$

☐ C. $150 \frac{m}{hr}$

☐ D. $200 \frac{m}{hr}$

☐ E. $250 \frac{m}{hr}$

41) Sam is considering a $2000 loan to purchase a bike. Which loan option results in the lowest interest payment?

☐ A. A 10-month loan with a 4.5% annual simple interest rate.

☐ B. A 15-month loan with a 4.25% annual simple interest rate.

☐ C. A 20-month loan with a 4% annual simple interest rate.

☐ D. A 24-month loan with a 3.75% annual simple interest rate.

☐ E. A 30-month loan with a 3.5% annual simple interest rate.

4.1 Practices

42) An arithmetic sequence is given by 2, 6, 10, 14, 18, ⋯. Which formula determines the nth term of this sequence?

- ☐ A. $3n+3$
- ☐ B. $4n-2$
- ☐ C. $4n+2$
- ☐ D. $5n-3$
- ☐ E. $5n+2$

43) A baker has 4 liters of milk in a jug. She uses 1.5 liters to make a batch of pastries. Then, she pours half of the remaining milk into a mixing bowl. Finally, she uses 500 milliliters of the remaining milk for a cake. How many milliliters of milk are left in the jug?

- ☐ A. 500 ml
- ☐ B. 750 ml
- ☐ C. 1000 ml
- ☐ D. 1250 ml
- ☐ E. 1500 ml

44) The following table represents the linear relationship between the revenue generated in dollars at a local bakery and the number of cakes sold. What is the rate of change in revenue with respect to the number of cake sales in the bakery?

Number of Cakes Sold	10	20	30
Revenue (in dollars)	200	400	600

- ☐ A. $10
- ☐ B. $15
- ☐ C. $20
- ☐ D. $25
- ☐ E. $30

45) Among the given tables, which one does not represent y as a function of x?

☐ A.

x	0	3	2	3	5
y	1	5	4	−5	−1

☐ B.

x	−3	−2	−1	0	1
y	0	1	2	−3	4

☐ C.

x	1	2	3	4	5
y	1	4	9	16	25

☐ D.

x	−2	−1	0	1	2
y	0	1	2	3	4

☐ E.

x	−4	−2	0	2	4
y	−2	−1	0	1	2

46) Simplify the expression $p^{\frac{3}{4}}q^{-1}p^3q^{\frac{1}{2}}$.

☐ A. $p^{\frac{15}{4}}q^{\frac{1}{2}}$

☐ B. $\dfrac{1}{p^{\frac{15}{4}}q^{\frac{1}{2}}}$

☐ C. $\dfrac{p^{\frac{15}{4}}}{q^{\frac{1}{2}}}$

☐ D. $\dfrac{p^{\frac{15}{2}}}{q^{\frac{1}{2}}}$

☐ E. $\dfrac{p^{\frac{3}{15}}}{q^{\frac{1}{2}}}$

47) Solve for y: $3 + \dfrac{4y}{y-6} = \dfrac{4}{6-y}$.

☐ A. $\dfrac{5}{6}$

☐ B. 2

☐ C. $\dfrac{9}{6}$

☐ D. 3

☐ E. $\dfrac{11}{6}$

48) The graph of $h(x) = x^2$ is transformed by a factor of 0.5 to make the graph of $j(x) = 0.5h(x)$. Which graphs show the transformation of h to j?

4.1 Practices

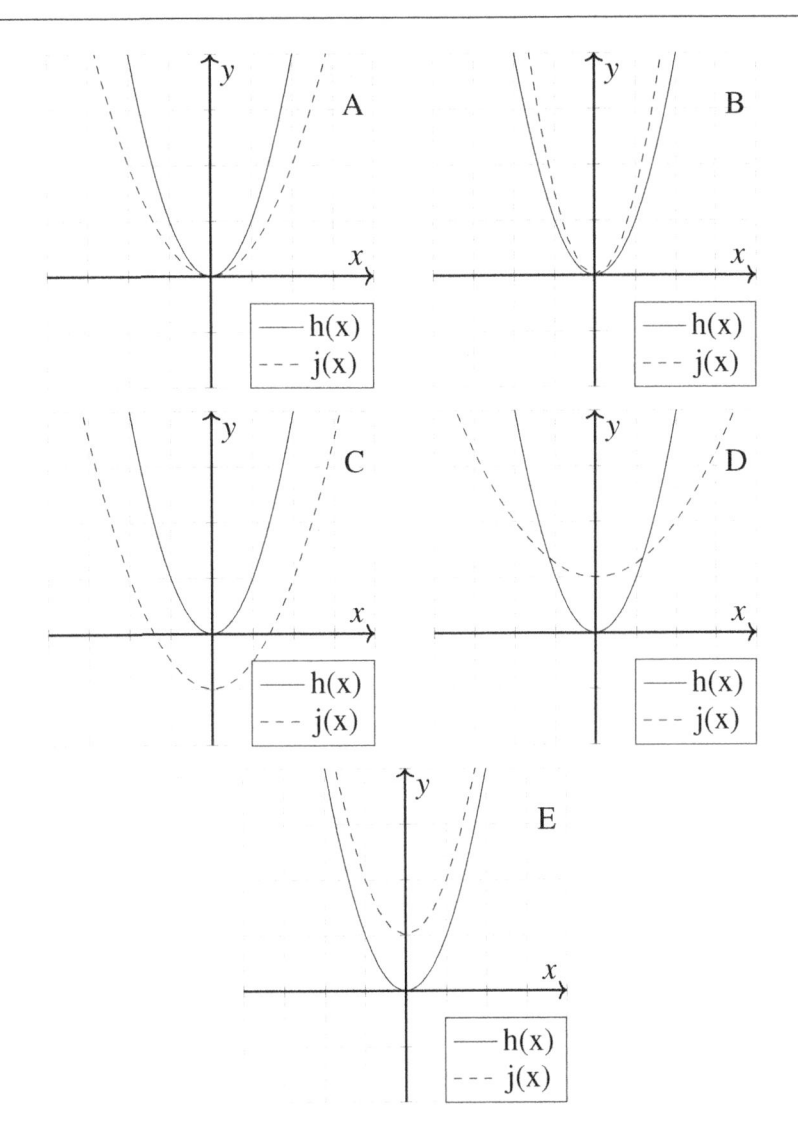

- ☐ A. Graph A
- ☐ B. Graph B
- ☐ C. Graph C
- ☐ D. Graph D
- ☐ E. Graph E

49) What is an equivalent expression for $4m^2 - 16$?

- ☐ A. $(2m-4)(2m-4)$
- ☐ B. $4(m-2)(m+2)$
- ☐ C. $4m(m-2)(m+2)$
- ☐ D. $4(m-4)$
- ☐ E. $4m(m-4)$

50) Determine the *y*-intercepts and *x*-intercepts of the equation $2x - 4y = 16$.

- ☐ A. $(0,-4), (8,0)$
- ☐ B. $(0,-4), (-8,0)$
- ☐ C. $(0,4), (-8,0)$
- ☐ D. $(0,4), (8,0)$
- ☐ E. $(0,-8), (4,0)$

4.2 Answer Keys

1) D. 11 units
2) B. $-\frac{11}{3}$
3) C. 508.68 m^3
4) A.
5) A. $y = \frac{1}{2}x$
6) C. $2500
7) A. $(x,y) \to (3+m, -4-n)$
8) C. 30
9) C. 60
10) D. 25
11) D. Graph D
12) C. Scaling by a factor of 0.5
13) C.
14) C. $\{(2,4), (3,6), (4,8), (5,10)\}$
15) C. 2
16) B. $V = \pi(2)^2(6)$
17) D. $(x,y) \to (0.8x, 0.8y)$
18) A. $y > 1\frac{9}{10}$
19) C. $\frac{8}{12}$
20) D. $2582.27
21) A. 25 cm
22) D. 80°
23) C. 2.5×10^3
24) C. $-2 < x \leq 1$
25) E. Graph E
26) C. $12
27) B. $3 per mile
28) D. $y = 135°$
29) C. 72
30) A. $(0, 7)$
31) C.
32) D. $y \geq 2x - 1$
33) C. $40y < 300 + 25y$
34) C. 60
35) C. $\frac{\overline{KL}}{\overline{JL}} = \frac{\overline{NO}}{\overline{MO}}$
36) A. $T = 4n + 15$
37) B. 142 m^2
38) D. 113.04 ft^3
39) C. 4
40) D. 200 $\frac{m}{hr}$
41) A. 10-month, 4.5%
42) B. $4n - 2$
43) B. 750 ml
44) C. $20
45) A.
46) C. $\frac{p^{\frac{15}{4}}}{q^{\frac{1}{2}}}$
47) B. 2
48) A. Graph A
49) B. $4(m-2)(m+2)$
50) A. $(0,-4), (8,0)$

4.3 Answers with Explanation

1) The distance between two points (x_1, y_1) and (x_2, y_2) is given by $d = \sqrt{(x_2 - x_1)^2 + (y_2 - y_1)^2}$. For points $Q(5, -4)$ and $S(-3, 3)$:

$$d = \sqrt{(-3-5)^2 + (3-(-4))^2} = \sqrt{(-8)^2 + 7^2} = \sqrt{64 + 49} = \sqrt{113} \approx 10.63 \text{ units.}$$

The closest value in the options is 11.

2) First, find the slope of the line using the points $(4, 3)$ and $(1, -2)$. The slope $m = \frac{y_2 - y_1}{x_2 - x_1} = \frac{-2-3}{1-4} = \frac{-5}{-3} = \frac{5}{3}$. Use the point-slope form to find the y-intercept: $y - y_1 = m(x - x_1)$. Using point $(4, 3)$, $y - 3 = \frac{5}{3}(x - 4)$. Setting $x = 0$ gives the y-intercept: $y = 3 - \frac{5}{3} \times 4 = 3 - \frac{20}{3} = -\frac{11}{3}$.

3) The formula for the volume of a cylinder is $V = \pi r^2 h$ (r for radius and h for height). For a tank with a radius of 3m and a height of 18m, the volume is $V = \pi \times 3^2 \times 18 = \pi \times 9 \times 18 = 162 \times 3.14 = 508.68$ m^3.

4) First, calculate the total interest for the simple interest account (3%):

$$\text{Total interest (Simple)} = \text{Principal} \times \text{Rate} \times \text{Time} = \$3000 \times 0.03 \times 5 = \$450.$$

Next, calculate the total amount for the compound interest account (2.8%):

$$\text{Total amount (Compound)} = \text{Principal} \times (1 + \text{Rate})^{\text{Time}} = \$3000 \times (1 + 0.028)^5 \approx \$3444.19.$$

Then subtract the principal to find the interest:

$$\text{Total interest (Compound)} = \text{Total amount (Compound)} - \text{Principal} = \$3444.19 - \$3000 \approx \$444.19.$$

Comparing the two:

$$\text{Interest difference} = \text{Total interest (Simple)} - \text{Total interest (Compound)} \approx \$450 - \$444.19 \approx 5.81.$$

5) A proportional relationship is one where the ratio of y to x is constant. Only option A, $y = \frac{1}{2}x$, represents a

4.3 Answers with Explanation

direct proportional relationship where the ratio of y to x is constant (i.e., 0.5 for every x).

6) Simple interest is calculated as $I = P \times r \times t$, where P is the principal, r is the rate, and t is time. For Julia's account: $I = 2000 \times 0.05 \times 5 = \$500$. The total balance is the original principal plus interest: $2000 + 500 = \$2500$.

7) Translation of a point (x,y) by m units right and n units down results in the new coordinates $(x+m, y-n)$. Therefore, the new center of the circle is at $(3+m, -4-n)$.

8) If 40% are proficient, then 60% are not proficient. For 50 students, 60% not proficient means $50 \times 0.60 = 30$ students are not proficient in the foreign language.

9) Mark's total grade sum is $6 \times 70 = 420$ using the formula average $= \frac{\text{sum of terms}}{\text{number of terms}}$. Since Emily has the same total, but for 7 courses, her average is $\frac{420}{7} = 60$.

10) To solve the problem, we use the equation: $90 - 5x = 40 + 5x$, where x represents the number of groups that Club Alpha gives to Club Beta. Simplifying, $10x = 50$, so $x = 5$. Since transfers occur in groups of 5, Club Alpha needs to transfer 5 groups of 5 students each to Club Beta, making a total of 25 students. This results in both clubs having 65 members each.

11) Michael painted 8 feet in 4 days, averaging 2 feet per day. The graph representing this rate should show a consistent increase of 2 feet per day. The starting point of the graph should be at zero because Michael started painting from the beginning. Therefore, the graph showing a steady slope of 2 feet per day from the origin correctly represents the progress.

12) A scaling transformation changes the size of a figure. A scale factor of 0.5 reduces the size of the original figure to half its original dimensions, thus altering its size but not its shape or orientation.

13) By examining the graph, we can estimate the costs for a 9-hour class at each school. The graph for School X shows a steeper slope compared to School Y, indicating a higher rate of increase in cost per hour. At 9 hours, the point on School X's graph is visibly higher than that on School Y's graph. The difference in heights, representing cost, appears to be around \$5. Therefore, the cost of a 9-hour class at School X is approximately \$5 more than that at School Y, aligning with option C.

14) A function has only one output for each input. Choice C is correct because each x value has a unique y

value, making it a function.

15) The y-intercept of a linear function ($y = mx + b$) is the constant term. For $y = 3x + 2$, the y-intercept is 2.

16) The volume of a cylinder is given by $V = \pi r^2 h$. The radius is half the diameter, so $r = \frac{4}{2} = 2$ inches. The height is 6 inches, so the volume formula is $V = \pi \times 2^2 \times 6$.

17) To shrink a figure, a dilation with a scale factor less than 1 is used. Choice D with a scale factor of 0.8 reduces the size of the square while maintaining its shape and orientation.

18) Since y is the length of the longer piece, it must be greater than half the total length of the ribbon. Half of $3\frac{4}{5}$ feet is $1\frac{9}{10}$ feet. Therefore, y must be greater than $1\frac{9}{10}$ feet.

19) The furmula of the Probability is: $\frac{\text{number of desired outcomes}}{\text{number of total outcomes}}$. The total number of balls is $4 + 6 + 2 = 12$. The number of non-orange balls is $6 + 2 = 8$. The probability of not picking an orange ball is $\frac{8}{12}$.

20) For Account X (simple interest): Balance = $P(1 + rt) = 1200(1 + 0.025 \times 3) = 1290$. For Account Y (compound interest): Balance = $P(1 + r)^t = 1200(1 + 0.025)^3 \approx 1292.27$. The sum of the balances is the total of both accounts' balances after 3 years: $1290 + 1292.27 = 2582.27$.

21) Using the Pythagorean theorem: $d^2 = 20^2 + 15^2$. Simplifying, $d^2 = 400 + 225 = 625$. Taking the square root, $d = \sqrt{625} = 25$ cm.

22) Let the smallest angle be $2x$. Then the angles are $2x, 3x, 4x$. The sum is $180°$: $2x + 3x + 4x = 180 \Rightarrow 9x = 180 \Rightarrow x = 20°$. The largest angle is $4x = 4 \times 20° = 80°$.

23) To convert 2500 to scientific notation, move the decimal point three places to the left. This gives 2.5×10^3.

24) The domain of a graph is the range of x-values it covers. Given the graph of an exponential function, the domain is determined by projecting the graph onto the x-axis. For this particular graph, the x-values start from the values bigger than -2 and extend to 1, hence the domain is $-2 < x \leq 1$.

25) The vertical line test is used to determine if a curve is a graph of a function. For a curve to represent a function, every vertical line drawn through the graph must intersect the curve at no more than one point. This ensures that for each input (x-value), there is exactly one output (y-value). Among the options:

Graph A, intersects a vertical line twice in certain areas, failing the test.

4.3 Answers with Explanation

Graph B, also intersects vertical lines more than once, indicating it is not a function.

Graph C, has two separate functions, one of them is below the other, failing the vertical line test.

Graph D, being a circle, intersects vertical lines at two points in most places.

Graph E, passes the vertical line test as every vertical line intersects the graph at exactly one point, satisfying the condition for a function.

Therefore, the correct answer is E, where the curve represents y as a function of x.

26) Translate the situation into an equation: $18 is $6 less than twice what she spent on a notebook. Let x denote the notebook's cost. The equation is $18 = 2x - 6$. Solve for x: $24 = 2x$, so $x = \frac{24}{2} = \$12$. Therefore, the notebook's price is $12.

27) To find the slope, use the formula: $m = \frac{y_2 - y_1}{x_2 - x_1}$. Choose two points, for instance, $(2, 10)$ and $(10, 34)$. The slope is calculated as $\frac{34-10}{10-2} = \frac{24}{8} = 3$. Therefore, the slope of the line representing the relationship between distance and fare is $3 per mile.

28) According to the triangle angle sum property, the sum of angles in a triangle is $180°$. Therefore, the third angle of the triangle is $45°$. Since y and the $45°$ angle of the triangle are adjacent angles forming a straight line, their sum is $180°$. Thus, $y = 180° - 45° = 135°$. Therefore, the correct answer is D.

29) Given that $\frac{a}{4} = b$ and $b = 2$, we can find the value of a by multiplying both sides of the equation by 4, yielding $a = 4 \times 2 = 8$. Now, we can calculate $a^2 + 4b$ as follows:

$$a^2 + 4b = 8^2 + 4 \times 2 = 64 + 8 = 72.$$

So, the correct answer is 72.

30) Testing each option in the equation:

A: $(0, 7)$ yields $5 \times 0 + 2 \times 7 = 14$, which matches.

B: $(2, 3)$ yields $5 \times 2 + 2 \times 3 = 16$, which does not match.

C: $(1, 4)$ yields $5 \times 1 + 2 \times 4 = 13$, which does not match.

D: $(3, 2)$ yields $5 \times 3 + 2 \times 2 = 19$, which does not match.

E: $(4, 1)$ yields $5 \times 4 + 2 \times 1 = 22$, which does not match.

Thus, the correct answer is A.

31) The inequality $7w < 5w+6$ implies that the total cost of 7 decorative pots ($7w$) is less than the combined cost of 5 pots and an additional $6 plant ($5w+6$). Therefore, buying 7 pots is more affordable than buying 5 pots along with a $6 plant, making option C the correct interpretation.

32) To determine the inequality for the shaded region, first identify the equation of the line. Two points on the line can be visually estimated from the graph, such as $(1,1)$ and $(2,3)$. Using these points, calculate the slope m of the line:
$$m = \frac{y_2 - y_1}{x_2 - x_1} = \frac{3-1}{2-1} = 2.$$
Since the line crosses the y-axis below the origin, the y-intercept b is negative. From the graph, $b = -1$. Thus, the equation of the line is $y = 2x - 1$. The shaded region includes points above this line. Therefore, the inequality representing this region is $y \geq 2x - 1$, corresponding to option D.

33) Jessica's total amount after y weeks is $\$300 + \$25y$, and Mark's total amount is $\$40y$. The condition for Jessica having more money is $\$300 + \$25y > \$40y$, which simplifies to $40y < 300 + 25y$.

34) Solve for c using the equation: $\frac{c}{5} = 3$. This gives $c = 15$. Now substitute $c = 15$ and $d = 3$ into $c + 5d^2$: $15 + 5 \times 3^2 = 15 + 5 \times 9 = 60$.

35) In right triangles, the slope is given by the ratio of the vertical side to the horizontal side. For triangle JKL, the slope is $\frac{KL}{JL}$ and for triangle MNO, it is $\frac{NO}{MO}$. Since these triangles are similar, their slopes are equal, resulting in the proportion $\frac{KL}{JL} = \frac{NO}{MO}$.

36) The total cost includes a fixed setup fee of $15 plus $4 per t-shirt. So, the function is $T = 4n + 15$, where n is the number of t-shirts.

37) The surface area of a cuboid is calculated using the formula: $2 \times (lw + lh + wh)$, where l is the length, w is the width, and h is the height of the cuboid. For this container, $l = 7$ m, $w = 5$ m, and $h = 3$ m. Plugging in these values, the surface area is $2 \times (7 \times 5 + 7 \times 3 + 5 \times 3) = 2 \times (35 + 21 + 15) = 2 \times 71 = 142$ m^2.

38) To calculate the volume of the balloon, use the formula for the volume of a sphere: $V = \frac{4}{3}\pi r^3$. The radius is half the diameter, so for 6 feet diameter, the radius is $r = 3$ feet. Substitute the radius into the formula: $V = \frac{4}{3} \times \pi \times 3^3$. Considering $\pi = 3.14$, this computes to $\frac{4}{3} \times 3.14 \times 27 = 113.04$ ft^3.

4.3 Answers with Explanation

39) To calculate the MAD, find the mean (average) score:

$$\frac{82+76+90+85+78+90+82+88+84+79}{10} = 83.4,$$

Then, calculate the absolute deviations from the mean for each score:

$$|82-83.4| = 1.4, \ |76-83.4| = 7.4, \ |90-83.4| = 6.6, \ |85-83.4| = 1.6, \ |78-83.4| = 5.4,$$

$$|90-83.4| = 6.6, \ |82-83.4| = 1.4, \ |88-83.4| = 4.6, \ |84-83.4| = 0.6, \ |79-83.4| = 4.4.$$

The mean of these absolute deviations is:

$$\text{MAD} = \frac{1.4+7.4+6.6+1.6+5.4+6.6+1.4+4.6+0.6+4.4}{10} = \frac{40}{10} = 4.$$

Therefore, the MAD is 4.

40) The rate of change, or the slope of the line in the graph, represents the distance covered per hour of battery power used. Using the points on the graph, for example, $(2,400)$ and $(4,800)$, the rate of change is calculated as $\frac{800-400}{4-2} = \frac{400}{2} = 200 \ \frac{m}{hr}$. This indicates that for each additional hour of battery power used, the drone covers 200 meters more, aligning with option D.

41) Use the simple interest formula $I = P \times r \times t$ to calculate the total interest for each loan option:

Option A: $I = 2000 \times 0.045 \times \frac{10}{12} = \$75$,

Option B: $I = 2000 \times 0.0425 \times \frac{15}{12} = \$106.25$,

Option C: $I = 2000 \times 0.04 \times \frac{20}{12} \approx \$133.3$,

Option D: $I = 2000 \times 0.0375 \times \frac{24}{12} = \$150$,

Option E: $I = 2000 \times 0.035 \times \frac{30}{12} = \$175$.

The option with the lowest interest payment is option A.

42) Use the formula for an arithmetic sequence $a_n = a_1 + (n-1)d$. Here, $a_1 = 2$ and the common difference $d = 4$. Thus, $a_n = 2 + (n-1) \times 4$. Simplifying, $a_n = 4n - 2$.

43) After using 1.5 liters, 2500 ml remain ($4\,l - 1.5\,l = 2.5\,l = 2500$ ml). She then pours half of the remaining milk into a bowl (2500 ml $\div 2 = 1250$ ml), leaving 2500 ml $- 1250$ ml $= 1250$ ml. After using 500 ml for the cake, 750 ml of milk are left in the jug (1250 ml − 500 ml = 750 ml).

44) Taking the points $(10, 200)$ and $(20, 400)$, the rate of change is $\frac{400-200}{20-10} = \frac{200}{10} = 20$. This means for every additional cake sold, the revenue increases by $20, corresponding to option C.

45) In a function, each input (or x value) should correspond to at most one output (or y value). In choice A, the x value of 3 is associated with two different y values (5 and -5), which violates the definition of a function. Therefore, choice A does not represent y as a function of x.

46) Simplify $p^{\frac{3}{4}}q^{-1}p^3q^{\frac{1}{2}}$ by combining like terms: $p^{\frac{3}{4}+3} = p^{\frac{15}{4}}$ and $q^{-1+\frac{1}{2}} = q^{-\frac{1}{2}}$. This simplifies to $\frac{p^{\frac{15}{4}}}{q^{\frac{1}{2}}}$.

47) Find a common denominator, $y-6$, and rewrite the equation: $3 + \frac{4y}{y-6} = \frac{3(y-6)}{y-6} + \frac{4y}{y-6} = \frac{3y-18+4y}{y-6} = \frac{7y-18}{y-6}$. Multiply $\frac{4}{6-y}$ by -1: $\frac{-4}{y-6}$. Set the numerators equal: $7y - 18 = -4$, solve for y: $7y = 14$, $y = \frac{14}{7} = 2$.

48) The function $j(x) = 0.5h(x)$ represents a vertical scaling of the original function $h(x) = x^2$ by a factor of 0.5. This means that the y-values of the graph of $h(x)$ are halved, resulting in a graph that is vertically shrinked compared to the original. In Graph A, the dashed curve (representing $j(x)$) is a vertically shrinked version of the solid curve (representing $h(x)$), with each y-value being half of the corresponding y-value on the original curve. This matches the description of the transformation, making Graph A the correct representation of the function $j(x) = 0.5h(x)$.

49) Apply the difference of squares formula, $a^2 - b^2 = (a+b)(a-b)$, to $4m^2 - 16$. This gives $4m^2 - 16 = (2m)^2 - 4^2 = (2m-4)(2m+4)$. It can also be written as: $2(m+2) \times 2(m-2) = 4(m-2)(m+2)$.

50) For the y-intercept, set $x = 0$: $-4y = 16 \Rightarrow y = -4$. The y-intercept is $(0, -4)$. To find the x-intercept, set $y = 0$: $2x = 16 \Rightarrow x = 8$. The x-intercept is $(8, 0)$.

5. Practice Test 4

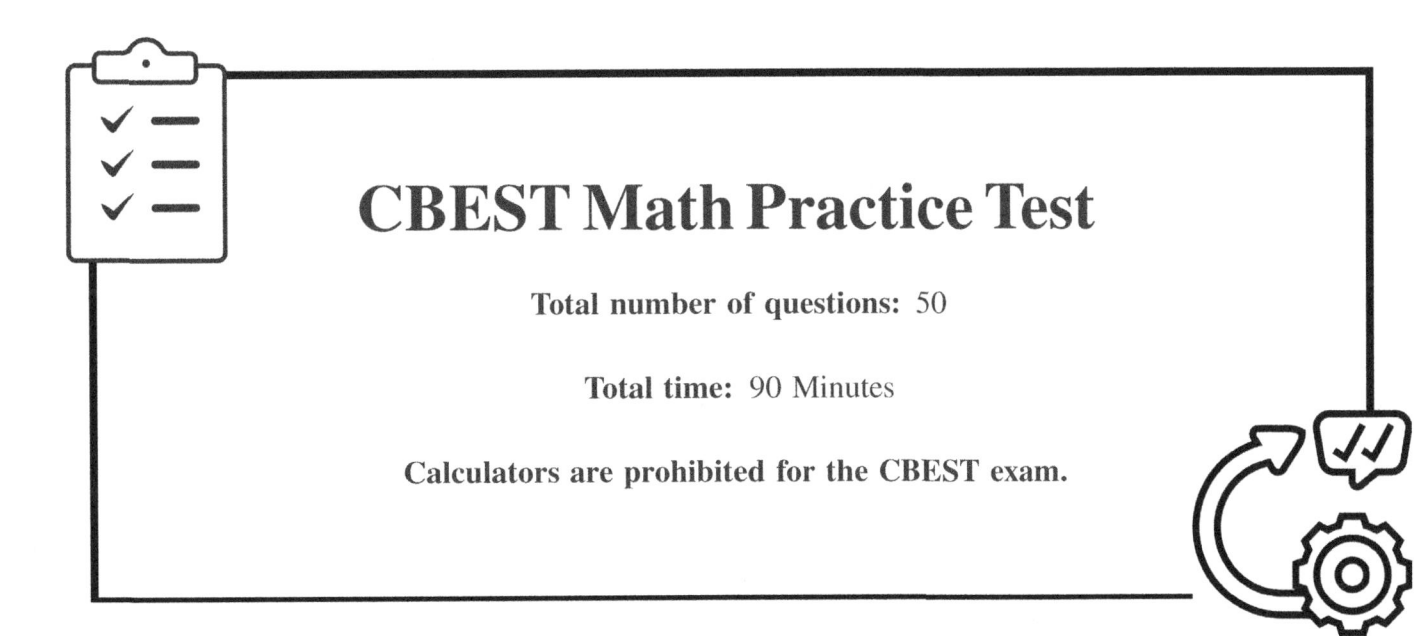

CBEST Math Practice Test

Total number of questions: 50

Total time: 90 Minutes

Calculators are prohibited for the CBEST exam.

5.1 Practices

1) A water tank in the shape of a cylinder has a radius of 5 centimeters and is 12 centimeters tall. What is the nearest value to its total surface area in square centimeters.

☐ A. 282.74 cm²
☐ B. 341.33 cm²
☐ C. 376.99 cm²
☐ D. 534.07 cm²

☐ E. 502.65 cm^2

2) Robert starts with $500 in his bank account and adds $25 weekly for z weeks. Meanwhile, Anna begins with $200 and saves $35 every week for the same duration. Identify the inequality representing when Robert's account exceeds Anna's.

☐ A. $25z < 500 + 35z$

☐ B. $25z > 500 + 35z$

☐ C. $35z < 300 + 25z$

☐ D. $35z < 300 - 25z$

☐ E. $35z > 500 - 25z$

3) Consider the net worth statement below for Oliver. Positive values indicate assets, while negative values show liabilities. The current value of Oliver's boat is missing.

Item	Value
Boat (current value)	
Savings	$2000
Credit card balance	−$3000
Stock portfolio	$20000
Home loan	−$150000
Pension fund	$60000
Auto loan	−$10000

If Oliver's net worth is $19000, what is the current value of his boat?

☐ A. $70000

☐ B. $82000

☐ C. $92000

☐ D. $100000

☐ E. $112000

4) A survey was conducted with 150 individuals regarding their most favored season. The following pie chart shows the proportion of participants favoring each season. How many more people prefer summer than winter?

5.1 Practices

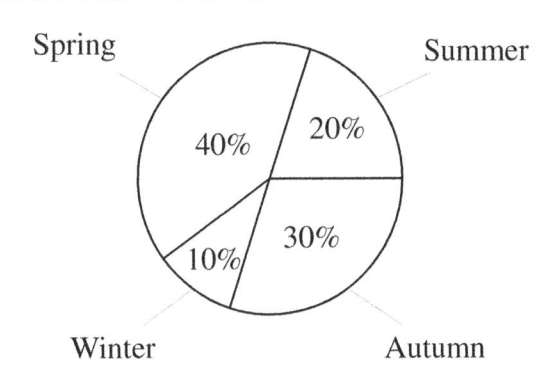

Spring 40%, Summer 20%, Autumn 30%, Winter 10%

- ☐ A. 10
- ☐ B. 15
- ☐ C. 20
- ☐ D. 25
- ☐ E. 30

5) A rectangular garden has a width of 8 meters, and the length of the diagonal is 17 meters. Determine the length of the garden in meters.

- ☐ A. 10 m
- ☐ B. 11 m
- ☐ C. 13 m
- ☐ D. 15 m
- ☐ E. 16 m

6) Mr. Thompson deposited $2000 into an account with no further transactions. The account earns 4.5% annual simple interest. Calculate the account balance after 2 years.

- ☐ A. $1890.00
- ☐ B. $2000.00
- ☐ C. $2090.00
- ☐ D. $2180.00
- ☐ E. $2270.00

7) A cylindrical grain silo has a diameter of 12 meters and a height of 20 meters. Which equation calculates V, the volume of the silo in cubic meters?

- ☐ A. $V = \pi(6)^2(20)$

- [ ] B. $V = \pi(6)^2(10)$
- [ ] C. $V = \pi(7)^2(15)$
- [ ] D. $V = \pi(12)^2(20)$
- [ ] E. $V = \pi(20)^2(6)$

8) The total cost of purchasing 15 books and 10 pencils is $65. Assuming that the price of each book and each pencil is the same, determine the cost of one book and one pencil.

- [ ] A. $1.50
- [ ] B. $2.60
- [ ] C. $3.00
- [ ] D. $4.20
- [ ] E. $5.00

9) A linear function crosses the point $(3,5)$ with a slope of $-\frac{1}{2}$. Determine the y-intercept of this function's graph.

- [ ] A. $\frac{1}{2}$
- [ ] B. $\frac{7}{2}$
- [ ] C. $\frac{5}{2}$
- [ ] D. $\frac{13}{2}$
- [ ] E. $\frac{9}{2}$

10) Two friends, Jake and Lara, take out loans from a bank.
- Jake borrows $5000 for 4 years with an annual simple interest rate of 6%.
- Lara borrows $5000 for 6 years with an annual simple interest rate of 5%.

What is the difference in the total interest paid by Jake and Lara?

- [ ] A. $300
- [ ] B. $350
- [ ] C. $400
- [ ] D. $450
- [ ] E. $500

11) Consider a triangular prism with a base having sides of lengths 7 cm, 9 cm, and 11 cm. The altitude of the

5.1 Practices

base triangle is 5 cm, and the prism's height is 20 cm. Determine the total surface area of this prism in square centimeters.

☐ A. 430 cm²

☐ B. 450 cm²

☐ C. 470 cm²

☐ D. 585 cm²

☐ E. 510 cm²

12) The mean of a set of 40 test scores was initially calculated to be 85. It was later discovered that one of the scores was incorrectly recorded as 90 when it should have been 65. What is the corrected mean of the test scores?

☐ A. 83

☐ B. 84

☐ C. 82.5

☐ D. 85

☐ E. 85.5

13) A graph represents the correlation between the height of a sunflower in feet and the number of days it has been growing. Which function best represents the relationship shown in the graph?

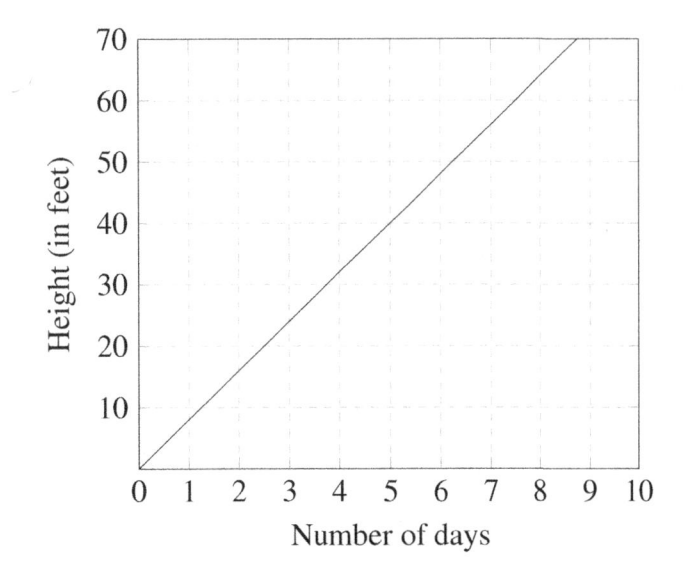

☐ A. $y = \frac{1}{8}x$

☐ B. $y = 2x$

- C. $y = 6x$
- D. $y = 8x$
- E. $y = 10x$

14) In a concert, the audience consists of adults and children with a ratio of 3 : 4. What could be the possible total number of people in the audience?
- A. 14036
- B. 21045
- C. 25047
- D. 28056
- E. 35071

15) Which sequence represents an arithmetic progression with a common difference of 3 and a first term of 1?
- A. 1, 4, 7, 10, 13, ...
- B. 1, 3, 5, 7, 9, ...
- C. 1, 2, 4, 7, 11, ...
- D. 1, 5, 9, 13, 17, ...
- E. 1, 3, 6, 9, 12, ...

16) Given the points $(2,4)$, $(4,8)$, $(6,12)$, $(8,16)$, and $(10,20)$, how would you describe this relationship?
- A. It represents y as a function of x because each x value corresponds to a unique y value.
- B. It does not represent y as a function of x, as each y value correlates to two x values.
- C. It represents y as a function of x because one y value corresponds to two x values.
- D. It does not represent y as a function of x, as each x value corresponds to two y values.
- E. The nature of the relationship cannot be determined.

17) Jessica has a monthly income of $8000. Her monthly expenses are detailed below. What percentage of her income is spent on rent and food?

Category	Amount
Rent	$2400
Food	$600
Other Expenses	$4000

5.1 Practices

- ☐ A. 25%
- ☐ B. 30%
- ☐ C. 37.5%
- ☐ D. 40%
- ☐ E. 45%

18) Consider a garden where the ratio of bees to flowers is 1 bees for every 3 flowers. Identify the graph that represents a relationship with the same unit rate as this scenario.

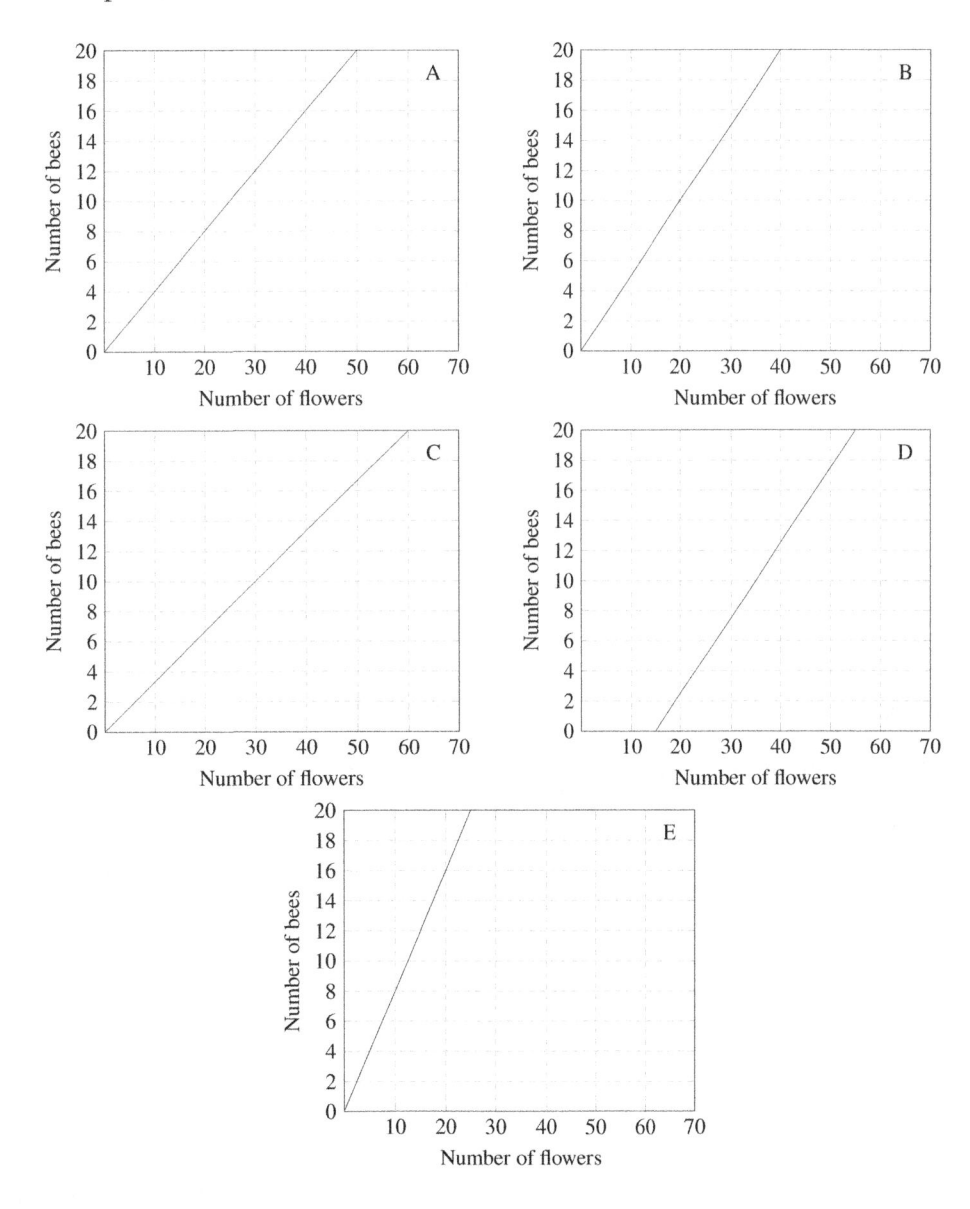

- ☐ A. Graph A
- ☐ B. Graph B

☐ C. Graph C

☐ D. Graph D

☐ E. Graph E

19) Alice has 40 marbles and Bob has 70 marbles. Alice gives away p marbles and Bob gives away q marbles. Afterwards, they have the same number of marbles. If Bob gave away three times as many marbles as Alice did, how many marbles did Alice give away?

☐ A. 6

☐ B. 8

☐ C. 10

☐ D. 12

☐ E. 15

20) Alex drove 180 miles in 4 hours, while Maria cycled 90 miles in 3 hours. What is the ratio of the average speed of Alex to that of Maria?

☐ A. 3 : 2

☐ B. 2 : 3

☐ C. 4 : 3

☐ D. 3 : 4

☐ E. 4 : 1

21) Consider a square graphed on a coordinate grid, with each side measuring 10 units. If this square is dilated by a scale factor of 0.5, with the origin as the center of dilation. What are the coordinates of a corresponding point on the resulting square if (x,y) is a point on the original square?

☐ A. $(0.5x, 0.5y)$

☐ B. $(0.6x, 0.6y)$

☐ C. $(0.8x, 0.8y)$

☐ D. $(1.0x, 1.0y)$

☐ E. $(2x, 2y)$

22) A company produces a product in two types of packaging: a cubical container and a cylindrical container. The cubical container has a side length of 8 cm, while the cylindrical container has a radius of 4 cm and a height of 12 cm. What is the difference in volumes of these two containers?

5.1 Practices

☐ A. The volume of the cubical container is about 64 cubic centimeters greater than the volume of the cylindrical container.

☐ B. The volume of the cubical container is about 100 cubic centimeters less than the volume of the cylindrical container.

☐ C. The volume of the cubical container is about 91 cubic centimeters less than the volume of the cylindrical container.

☐ D. The volume of the cubical container is about 200 cubic centimeters less than the volume of the cylindrical container.

☐ E. The volume of the cubical container is about 256 cubic centimeters greater than the volume of the cylindrical container.

23) A landscaper is designing a triangular flower bed. The lengths of two sides of the triangle are given, and these sides meet at a right angle. The landscaper wants to calculate the length of the third side. Which of the following sets of side lengths does NOT form a right triangle?

☐ A. 7 m, 24 m, 25 m

☐ B. 9 m, 40 m, 41 m

☐ C. 12 m, 35 m, 38 m

☐ D. 8 m, 15 m, 17 m

☐ E. 11 m, 60 m, 61 m

24) The table shows the linear relationship between the number of days, d, and the number of miles Sara jogged, m.

Number of Days (d)	2	5	8	10
Number of Miles Jogged (m)	10	25	40	50

Based on the table, how many miles does Sara jog per day?

☐ A. 2 miles per day

☐ B. 4 miles per day

☐ C. 5 miles per day

☐ D. 6 miles per day

☐ E. 7 miles per day

25) A garden has 20 trees with an average height of 10 feet. After removing the five tallest trees, which have an average height of 15 feet, what is the nearest value to the average height of the remaining trees?

- ☐ A. 5 feet
- ☐ B. 7 feet
- ☐ C. 8 feet
- ☐ D. 9 feet
- ☐ E. 10 feet

26) A bag contains 3 red marbles, 4 blue marbles, and 5 green marbles. If a single marble is picked at random, what is the probability that it is either red or blue?

- ☐ A. $\frac{3}{12}$
- ☐ B. $\frac{7}{12}$
- ☐ C. $\frac{5}{12}$
- ☐ D. $\frac{4}{12}$
- ☐ E. $\frac{1}{2}$

27) If $c = 3$ and $\frac{d}{5} = c$, what is the value of $d^2 + 5c$?

- ☐ A. 84
- ☐ B. 240
- ☐ C. 96
- ☐ D. 104
- ☐ E. 209

28) A cyclist starts from point A and rides 10 miles north. After a short break, the cyclist then rides 15 miles east. What is the nearest value to the straight-line distance from the starting point to the cyclist's final position?

- ☐ A. 18 miles
- ☐ B. 20 miles
- ☐ C. 22 miles
- ☐ D. 25 miles
- ☐ E. 30 miles

29) A student needs to save for a 6-month language course costing $900. They have already saved $300. How much does the student need to save each month to cover the remaining cost?

5.1 Practices

☐ A. $50

☐ B. $75

☐ C. $100

☐ D. $125

☐ E. $150

30) A car travels to four cities, covering an average distance of 150 miles between each city. If the distance to a fifth city, more than 250 miles away, is added, which of the following could be the new average distance?

☐ A. 150 miles

☐ B. 155 miles

☐ C. 160 miles

☐ D. 165 miles

☐ E. 170 miles

31) Express the number 0.000067890 in scientific notation.

☐ A. 6.7890×10^{-5}

☐ B. 6.7890×10^{-4}

☐ C. 67.890×10^{5}

☐ D. 67.890×10^{-8}

☐ E. 6.7890×10^{5}

32) Find the number for which 60% of its value is equal to 12.

☐ A. 12

☐ B. 18

☐ C. 20

☐ D. 22

☐ E. 24

33) Mark completed 4 laps in a swimming pool in 60 seconds. Which of the following represents an equivalent rate of swimming?

☐ A. 3 laps in 50 seconds

☐ B. 5 laps in 75 seconds

- [ ] C. 6 laps in 80 seconds
- [ ] D. 7 laps in 100 seconds
- [ ] E. 8 laps in 130 seconds

34) Lisa and Meg jointly babysat for 8 hours and earned $120 together. If they shared the earnings equally and Lisa spent $30 on a new game, how much money did she have left?
- [ ] A. $30
- [ ] B. $40
- [ ] C. $50
- [ ] D. $60
- [ ] E. $70

35) What is the closest option to the volume of a sphere with a radius of approximately 3 inches?
- [ ] A. 36 in^3
- [ ] B. 50 in^3
- [ ] C. 113 in^3
- [ ] D. 128 in^3
- [ ] E. 150 in^3

36) Determine the value of x based on the given model.

$$\boxed{x}\;\boxed{x}\;\boxed{x}\;\boxed{x}\;\boxed{x}\;\;\boxed{1}\;\boxed{1}\;\;=\;\;\boxed{x}\;\boxed{x}\;\;\boxed{1}\;\boxed{1}\;\boxed{1}$$

- [ ] A. $x = \frac{2}{3}$
- [ ] B. $x = -\frac{1}{3}$
- [ ] C. $x = -\frac{2}{7}$
- [ ] D. $x = -\frac{1}{7}$
- [ ] E. $x = \frac{1}{3}$

37) Sarah has a computer monitor with a screen that measures 24 inches diagonally and has an aspect ratio of 16:9. She plans to buy a new monitor that is larger, maintaining the same aspect ratio but increasing the diagonal size by a scale factor of 1.2. What will be the diagonal size, d, of Sarah's new monitor?

5.1 Practices

☐ A. $d = 28.8$ inches

☐ B. $d = 29.4$ inches

☐ C. $d = 30.2$ inches

☐ D. $d = 31.0$ inches

☐ E. $d = 32.5$ inches

38) Kevin documented his reading progress over two weeks. He read 600 pages of a novel and 400 pages of a science book. Which proportion can be used to determine q, the percentage of the total number of pages read that were from the novel?

☐ A. $\frac{q}{100} = \frac{600}{1000}$

☐ B. $\frac{q}{100} = \frac{400}{1000}$

☐ C. $\frac{q}{100} = \frac{1000}{600}$

☐ D. $\frac{q}{100} = \frac{600}{400}$

☐ E. $\frac{q}{100} = \frac{400}{600}$

39) In a cooking competition, three chefs were asked to prepare dishes with a specific calorie content. Their dishes' calorie contents were:

Chef 1: 250 Calories

Chef 2: $275\frac{3}{5}$ Calories

Chef 3: $260\sqrt{2}$ Calories

Which list shows these calorie contents in order from greatest to least?

☐ A. 250 Calories, $260\sqrt{2}$ Calories, $275\frac{3}{5}$ Calories

☐ B. 250 Calories, $275\frac{3}{5}$ Calories, $260\sqrt{2}$ Calories

☐ C. $260\sqrt{2}$ Calories, 250 Calories, $275\frac{3}{5}$ Calories

☐ D. $260\sqrt{2}$ Calories, $275\frac{3}{5}$ Calories, 250 Calories

☐ E. $275\frac{3}{5}$ Calories, 250 Calories, $260\sqrt{2}$ Calories

40) An astronomer observes a distant star and estimates its distance from Earth to be 0.0000045 light years. what is the scientific notation of this number?

☐ A. 4.5×10^{-3} light years

☐ B. 4.5×10^{-4} light years

☐ C. 4.5×10^{-5} light years

- [ ] D. 4.5×10^{-6} light years
- [ ] E. 4.5×10^{-7} light years

41) What is the probability of rolling a number greater than 4 on a standard six-sided die?
- [ ] A. $\frac{1}{6}$
- [ ] B. $\frac{1}{3}$
- [ ] C. $\frac{1}{2}$
- [ ] D. $\frac{2}{3}$
- [ ] E. $\frac{5}{6}$

42) On a coordinate grid, a triangle *DEF* is defined. Triangle *DEF* undergoes a contraction with a scale factor of *r*, centered at the origin, resulting in triangle *D'E'F'*. Which ordered pair accurately represents the coordinates of point *F'*?

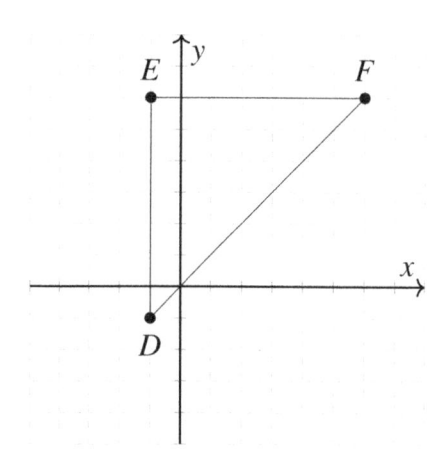

- [ ] A. $(\frac{6}{r}, \frac{6}{r})$
- [ ] B. $(6r, 6r)$
- [ ] C. $(r-6, r-6)$
- [ ] D. $(6+r, 6+r)$
- [ ] E. $(6-r, 6-r)$

43) Which numbers from the set $\{-6, -2, 0, 4, 7, 9\}$ satisfy the inequality $5 - 2y > 3$?
- [ ] A. $-6, -2$
- [ ] B. $-6, -2, 0$
- [ ] C. $4, 7, 9$

5.1 Practices

- [ ] D. $-6, -2, 0, 4$
- [ ] E. $0, 4, 7, 9$

44) In a coordinate plane, a rectangle $EFGH$ is rotated $90°$ clockwise around the origin to form rectangle $E'F'G'H'$.

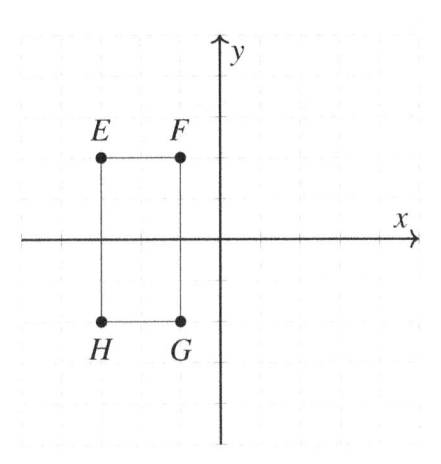

Which of the following statements is true?

- [ ] A. Rectangle $EFGH$ is larger than rectangle $E'F'G'H'$.
- [ ] B. The perimeter of rectangle $EFGH$ is different from the perimeter of rectangle $E'F'G'H'$.
- [ ] C. The angles of rectangle $EFGH$ are more acute than those of rectangle $E'F'G'H'$.
- [ ] D. The area of rectangle $EFGH$ is the same as the area of rectangle $E'F'G'H'$.
- [ ] E. The diagonal lengths of rectangle $EFGH$ are shorter than those of rectangle $E'F'G'H'$.

45) A bakery produces and sells boxes of cookies. The bakery incurs fixed costs of $30000, and it costs an additional $5 to make each box of cookies. The selling price for each box is $15. The graph of the system of linear equations representing the bakery's costs and revenue for manufacturing and selling x boxes of cookies is shown below. What is the number of boxes of cookies the bakery needs to sell to achieve equal costs and revenue?

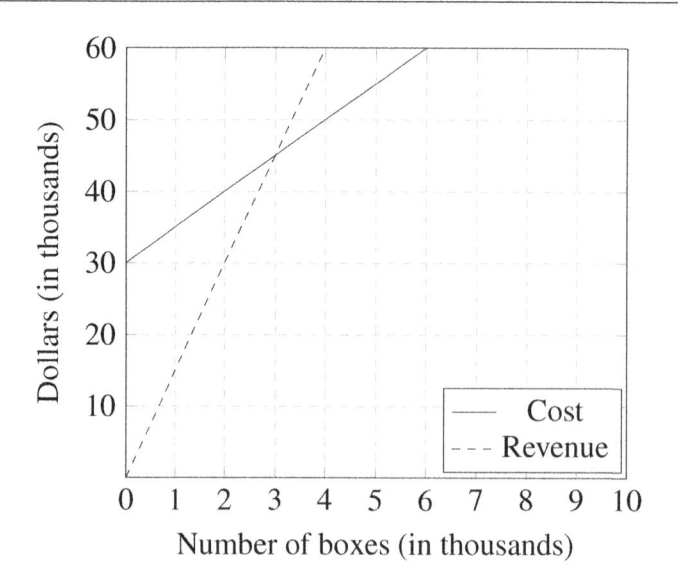

☐ A. 2000

☐ B. 2500

☐ C. 3000

☐ D. 4000

☐ E. 5000

46) Consider a parabola in the xy-plane described by the equation $h(x) = x^2 - 10x + 21$. This parabola intersects the x-axis at two points. Determine the distance between these two points of intersection.

☐ A. 1

☐ B. 2

☐ C. 3

☐ D. 4

☐ E. 5

47) A polynomial function $p(x)$ is given. Determine which of the following must be a factor of $p(x)$ based on the values in the table below.

x	-2	0	2	3
$p(x)$	0	-3	1	-10

☐ A. $x - 3$

☐ B. $x + 2$

☐ C. $x - 2$

5.1 Practices

☐ D. x

☐ E. $x+3$

48) The graph of a linear equation in the form $y = mx + b$ is represented on the grid. Based on the graph, what is the value of x when $y = 6$?

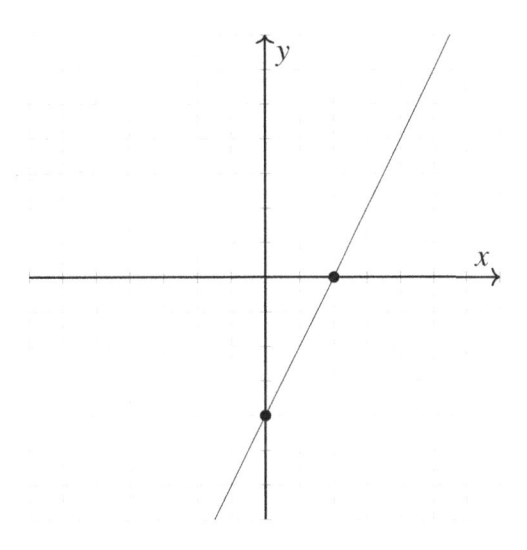

☐ A. -5

☐ B. -3

☐ C. 3

☐ D. 5

☐ E. 7

49) An online bookstore sells hardcover books for $15.00 each and paperback books for $10.00 each. If the store sold a total of 200 books, generating a revenue between $2500 and $3000, which inequality represents the possible number of hardcover books sold?

☐ A. $50 \leq h < 100$

☐ B. $50 < h \leq 150$

☐ C. $50 \leq h \leq 180$

☐ D. $100 < h < 200$

☐ E. $100 \leq h \leq 180$

50) The graph of $h(x) = x^2$ is plotted on a coordinate grid. What is true about the relationship between the graph of h and the graph of $k(x) = (2x)^2$?

☐ A. The graph of k is narrower than the graph of h

☐ B. The graph of k is flatter than the graph of h.

☐ C. The graph of k is shifted 2 units right compared to the graph of h.

☐ D. The graph of k is shifted 2 units down from the graph of h.

☐ E. The graph of k is shifted 2 units up from the graph of h.

5.2 Answer Keys

1) D. 534.07 cm^2
2) C. $35z < 300 + 25z$
3) D. $100000
4) B. 15
5) D. 15 m
6) D. $2180.00
7) A. $V = \pi(6)^2(20)$
8) B. $2.60
9) D. $\frac{13}{2}$
10) A. $300
11) D. 585 cm^2
12) B. 84
13) D. $y = 8x$
14) D. 28056
15) A. 1, 4, 7, 10, 13, ...
16) A. It represents y as a function of x because each x value corresponds to a unique y value.
17) C. 37.5%
18) C. Graph C
19) E. 15
20) A. 3 : 2
21) A. $(0.5x, 0.5y)$
22) C.
23) C. 12 m, 35 m, 38 m
24) C. 5 miles per day
25) C. 8 feet
26) B. $\frac{7}{12}$
27) B. 240
28) A. 18 miles
29) C. $100
30) E. 170 miles
31) A. 6.7890×10^{-5}
32) C. 20
33) B. 5 laps in 75 seconds
34) A. $30
35) C. 113 in^3
36) E. $x = \frac{1}{3}$
37) A. $d = 28.8$ inches
38) A. $\frac{q}{100} = \frac{600}{1000}$
39) D. $260\sqrt{2}$ Calories, $275\frac{3}{5}$ Calories, 250 Calories
40) D. 4.5×10^{-6} light years
41) B. $\frac{1}{3}$
42) A. $(\frac{6}{r}, \frac{6}{r})$
43) B. $-6, -2, 0$
44) D.
45) C. 3000
46) D. 4
47) B. $x + 2$
48) D. 5
49) D. $100 < h < 200$
50) A. k is narrower than h.

5.3 Answers with Explanation

1) The total surface area of a cylinder is given by $2\pi r^2 + 2\pi rh = 2\pi r(r+h)$. Here, $r = 5$ cm and $h = 12$ cm. Therefore, the surface area is $2\pi \times 5 \times (5+12) = 2\pi \times 5 \times 17 = 170\pi = 534.07$ cm^2.

2) To determine when Robert's savings exceed Anna's, we compare their respective account balances. Robert starts with $500 and adds $25 each week for z weeks, leading to his total savings being $500 + 25z$. Meanwhile, Anna begins with $200 and saves $35 each week for the same z weeks, accumulating a total of $200 + 35z$. The question asks when Robert's account exceeds Anna's. This translates to finding when $500 + 25z$ (Robert's savings) is greater than $200 + 35z$ (Anna's savings). Simplifying this inequality:

$$500 + 25z > 200 + 35z \Rightarrow 300 + 25z > 35z.$$

3) Net worth is calculated as Assets minus Liabilities. Oliver's net worth is $19000, comprising his savings, stock portfolio, pension fund, and the value of the boat, minus his debts. Thus,

$$\$19000 = \text{Boat} + \$2000 + \$20000 + \$60000 - \$3000 - \$150000 - \$10000.$$

Solving for the Boat's value gives: Boat = $100000.

4) Number of people surveyed = 150. Percentage of people who prefer summer = 20%. Therefore, the number of people who prefer summer is: $20\% \times 150 = 0.20 \times 150 = 30$. Percentage of people who prefer winter = 10%. Therefore, the number of people who prefer winter is: $10\% \times 150 = 0.10 \times 150 = 15$. The difference in numbers is $30 - 15 = 15$.

5) Using the Pythagorean theorem, where the diagonal is the hypotenuse, we have $l^2 + 8^2 = 17^2$. Thus, $l^2 = 17^2 - 8^2 = 289 - 64 = 225$. Therefore, $l = \sqrt{225} = 15\ m$.

6) Using the simple interest formula $I = Prt$, where $P = \$2000$, $r = 0.045$, and $t = 2$. So, $I = \$2000 \times 0.045 \times 2 = \$180$. The final balance is $P + I = \$2000 + \$180 = \$2180.00$.

7) The volume of a cylinder is calculated as $\pi r^2 h$. Here, the radius is half the diameter, so $r = 6$ meters, and

5.3 Answers with Explanation

the height is 20 meters. Therefore, $V = \pi(6)^2(20)$.

8) Let the cost of each book or pencil be y. The total cost for 15 books and 10 pencils is $15y + 10y = \$65$. Simplifying, $25y = \$65$, we find $y = \$2.60$.

9) The equation of a line is $y = mx + b$, where m is the slope and b is the y-intercept. Using the slope $m = -\frac{1}{2}$ and point $(3,5)$, we find $5 = -\frac{1}{2}(3) + b$. Solving for b, we get $b = \frac{13}{2}$.

10) For Jake, using $I = Prt$, $I = \$5000 \times 0.06 \times 4 = \$1200$. For Lara, $I = \$5000 \times 0.05 \times 6 = \$1500$. The difference is $\$1500 - \$1200 = \$300$.

11)

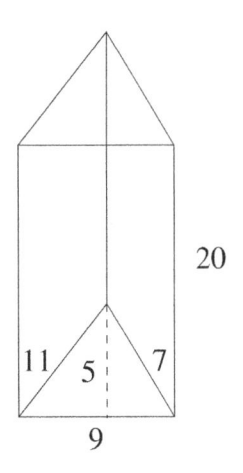

The surface area of the prism is: $A = 2B + Ph$, where A is the total surface area, B is the area of the base, P is the perimeter of the base, and h is the height of the prism. The base area is $\frac{1}{2} \times 9 \text{ cm} \times 5 \text{ cm} = 22.5 \text{ cm}^2$. The perimeter, $P = 7 \text{ cm} + 9 \text{ cm} + 11 \text{ cm} = 27 \text{ cm}$. Thus, $A = 2 \times 22.5 \text{ cm}^2 + (27 \text{ cm} \times 20 \text{ cm}) = 45 \text{ cm}^2 + 540 \text{ cm}^2 = 585 \text{ cm}^2$.

12) Initially, mean $= \frac{\text{sum of terms}}{\text{number of terms}} = \frac{\text{sum of terms}}{40} = 85$, so we have: sum of terms $= 85 \times 40 = 3400$. The difference between the misread and actual scores is $90 - 65 = 25$. The corrected sum $= 3400 - 25 = 3375$. Thus, corrected mean $= \frac{3375}{40} = 84.375$, approximately 84.

13) To determine the relationship shown in the graph, calculate the slope. Select two points on the graph, for example: $(5,40)$ and $(0,0)$. Calculate the slope using the formula: $m = \frac{y_2 - y_1}{x_2 - x_1} \Rightarrow m = \frac{40-0}{5-0} \Rightarrow m = 8$. To write the equation for this graph, use $y = mx + b$. The y-intercept (b) is found by substituting the values: $y = mx + b \Rightarrow 0 = 8 \times 0 + b \Rightarrow b = 0$. Therefore, the equation is $y = 8x$, making the correct answer option D.

14) The total number of people must be divisible by $3+4=7$. Only 28056 is divisible by 7 without a remainder.

15) Starting with 1, each subsequent term is obtained by adding the common difference of 3: $1+3=4$, $4+3=7$, and so on. Thus, the sequence is 1, 4, 7, 10, 13, ...

16) The relation is a function because for each x value, there is exactly one corresponding y value, fulfilling the definition of a function.

17) The total spent on rent and food is $\$2400 + \$600 = \$3000$. The percentage of income used for these expenses is $\frac{\$3000}{\$8000} \times 100\% = 37.5\%$.

18) Given the ratio of bees to flowers is 1 bee for every 3 flowers, the slope of the graph representing this relationship is $\frac{1}{3}$. This ratio means that for every 3 flowers, there is 1 bee, making the relationship directly proportional. In a graph where the number of bees and flowers are dependent on each other, if there are no flowers, there would be no bees. Therefore, the graph must start at the origin (0,0), indicating zero bees when there are zero flowers. The graph should maintain a constant slope of $\frac{1}{3}$. This means the line should increase by 1 unit on the y-axis (number of bees) for every 3 units increase on the x-axis (number of flowers). Evaluating the given graphs, Graph C is the only one that starts at the origin and has a constant slope of $\frac{1}{3}$, matching the ratio of bees to flowers. Thus, the correct answer is option C.

19) Let p be the number of marbles Alice gave away. Then Bob gave away $3p$ marbles. The equation is $40 - p = 70 - 3p$. Solving for p gives $p = 15$.

20) Alex's average speed is $\frac{180}{4} = 45$ miles per hour and Maria's is $\frac{90}{3} = 30$ miles per hour. The ratio is $45 : 30$, which simplifies to $3 : 2$.

21) With a scale factor of 0.5, each coordinate is halved. Thus, the coordinates of the corresponding point on the resulting square are $(0.5x, 0.5y)$.

22) Volume of cubical container: $8^3 = 512$ cm^3. Volume of cylindrical container: $\pi \times 4^2 \times 12 \approx 603$ cm^3. Difference: $603 - 512 = 91$ cm^3. The cubic container's volume is approximately 91 cubic centimeters less than that of the cylindrical container.

23) To verify if a set of side lengths forms a right triangle, we use the Pythagorean theorem: $a^2 + b^2 = c^2$,

5.3 Answers with Explanation

where c is the hypotenuse.

Option A: $7^2 + 24^2 = 49 + 576 = 625$, which equals 25^2. This supports a right triangle.

Option B: $9^2 + 40^2 = 81 + 1600 = 1681$, which equals 41^2. This supports a right triangle.

Option C: $12^2 + 35^2 = 144 + 1225 = 1369$, which does not equal 38^2 (1444). This does not support a right triangle.

Option D: $8^2 + 15^2 = 64 + 225 = 289$, which equals 17^2. This supports a right triangle.

Option E: $11^2 + 60^2 = 121 + 3600 = 3721$, which equals 61^2. This supports a right triangle.

Therefore, option C is the correct answer.

24) The ratio of the number of days to the number of miles Sara jogged is 2 to 10. Thus, the rate is $\frac{10}{2} = 5$ miles per day.

25) Total initial height: $10 \times 20 = 200$ feet. Total height of tallest trees: $15 \times 5 = 75$ feet. Height of remaining trees: $200 - 75 = 125$ feet. Average height of remaining trees: $\frac{125}{15} \approx 8.3$ feet, which rounds to 8 feet.

26) The total number of marbles is $3 + 4 + 5 = 12$. The number of red and blue marbles is $3 + 4 = 7$. Therefore, the probability of picking a red or blue marble is $\frac{7}{12}$.

27) Given $\frac{d}{5} = c$ and $c = 3$, we find the value of d by multiplying both sides of the equation by 5, which gives $d = 5 \times 3 = 15$. Now, we calculate $d^2 + 5c$ as follows:

$$d^2 + 5c = 15^2 + 5 \times 3 = 225 + 15 = 240.$$

Therefore, the correct answer is 240.

28) The problem can be solved using the Pythagorean theorem. The distance traveled north (10 miles) and east (15 miles) form the legs of a right triangle. The straight-line distance is the hypotenuse, calculated as $\sqrt{10^2 + 15^2} = \sqrt{100 + 225} = \sqrt{325} \approx 18$ miles.

29) The student needs an additional $900 - 300 = 600$. To save this amount over 6 months, the student needs to save $600 \div 6 = 100$ per month.

30) Initial total distance for four cities: $150 \times 4 = 600$ miles. With the addition of a fifth city more than 250 miles away, the new total distance is more than 850 miles. New average distance: more than $850 \div 5 = 170$

miles.

31) To express the number in scientific notation, the number is written in the form $m \times 10^n$, where m is between 1 and 10. For the number 0.000067890, move the decimal point to the right until you get a number between 1 and 10, resulting in 6.7890. Count the number of places the decimal moved: it moved 5 places. Therefore, the power of 10 is -5. Remember, when the decimal moves to the right, the exponent is negative. Thus, $0.000067890 = 6.7890 \times 10^{-5}$.

32) Let the number be z. Since 60% of z is 12, we have $0.60z = 12$. Solving for z gives $z = 12 \div 0.60 = 20$.

33) Mark's rate is $\frac{4 \text{ laps}}{60 \text{ seconds}} = \frac{1 \text{ lap}}{15 \text{ seconds}}$. Option B, 5 laps in 75 seconds, is equivalent as $\frac{5 \text{ laps}}{75 \text{ seconds}} = \frac{1 \text{ lap}}{15 \text{ seconds}}$.

34) Total earnings: $120. Lisa's share: $\frac{\$120}{2} = \$60$. After spending $30, she has $60 - \$30 = \$30$ left.

35) Volume of a sphere: $V = \frac{4}{3}\pi r^3$. With $r = 3$ inches, $V \approx \frac{4}{3}\pi(3)^3 \approx 113$ in^3.

36) To determine the value of x, count the number of x-tiles and 1-tiles on each side of the equation. On the left side, there are 5 x-tiles and 2 1-tiles. On the right side, there are 2 x-tiles and 3 1-tiles. Form the equation based on this model:

$$5x + 2 = 2x + 3.$$

To solve for x, first, get all the x-terms on one side and the numerical terms on the other side: $5x - 2x = 3 - 2 \Rightarrow 3x = 1$. Now, divide both sides by 3 to isolate x: $x = \frac{1}{3}$. Therefore, the correct value of x is $\frac{1}{3}$, which corresponds to choice E.

37) By increasing the diagonal size by a factor of 1.2, the new diagonal size is: $24 \text{ in} \times 1.2 = 28.8$ in.

38) To determine q, the percentage of total pages read from the novel, we set up a fraction comparing the number of pages read from the novel to the total pages read. Kevin read 600 pages of the novel out of a total of 1000 pages (600 from the novel + 400 from the science book). Thus, the proportion is $\frac{q}{100} = \frac{600}{1000}$.

39) Converting $260\sqrt{2}$ Calories to a decimal, it is approximately 367.69 Calories. Also $275\frac{3}{5} = \frac{1378}{5}$ is equal to 275.6. So, the order from greatest to least is: 367.69 Calories ($260\sqrt{2}$ Calories), 275.6 Calories, 250 Calories.

40) To convert 0.0000045 to scientific notation, we need to express it in the form of $a \times 10^n$, where $1 \leq |a| < 10$

5.3 Answers with Explanation

and n is an integer. In 0.0000045, the decimal point is moved 6 places to the right to get the number 4.5. Therefore, the exponent in the power of 10 becomes -6. This means the scientific notation of 0.0000045 light years is 4.5×10^{-6} light years.

41) On a standard six-sided die, the numbers greater than 4 are 5 and 6. There are two favorable outcomes out of six possible outcomes, so the probability is $\frac{2}{6} = \frac{1}{3}$.

42) When Triangle DEF is contracted by a scale factor of r from the origin, each point of the triangle is scaled by $\frac{1}{r}$. The original location of point F is at $(6,6)$. To find the new location of point F', we apply the scale factor to F's coordinates: $\left(\frac{1}{r} \times 6, \frac{1}{r} \times 6\right) = \left(\frac{6}{r}, \frac{6}{r}\right)$. Thus, the coordinates of point F' after the contraction are $\left(\frac{6}{r}, \frac{6}{r}\right)$, making option A the correct choice.

43) Test each number in the set $\{-6, -2, 0, 4, 7, 9\}$ to see if it satisfies the inequality $5 - 2y > 3$:
- $y = -6 \rightarrow 5 - 2(-6) > 3 \rightarrow 17 > 3$, true.
- $y = -2 \rightarrow 5 - 2(-2) > 3 \rightarrow 9 > 3$, true.
- $y = 0 \rightarrow 5 - 2(0) > 3 \rightarrow 5 > 3$, true.
- $y = 4 \rightarrow 5 - 2(4) > 3 \rightarrow -3 > 3$, false.
- $y = 7 \rightarrow 5 - 2(7) > 3 \rightarrow -9 > 3$, false.
- $y = 9 \rightarrow 5 - 2(9) > 3 \rightarrow -13 > 3$, false.

The values that satisfy the inequality are $-6, -2, 0$. Thus, the correct answer is option B.

44) Analyzing each statement:

Option A: Rotation does not change the size of a shape, so $EFGH$ and $E'F'G'H'$ are the same size. This statement is false.

Option B: The perimeter remains unchanged during rotation. This statement is false.

Option C: The angle measures in a rectangle are always $90°$, regardless of rotation. This statement is false.

Option D: The area of a shape remains constant during rotation. Since the dimensions of $EFGH$ and $E'F'G'H'$ are the same, their areas are equal. This statement is true.

Option E: The diagonal lengths of a rectangle are invariant under rotation. This statement is false.

45) To find the number of boxes of cookies that the bakery needs to sell to achieve equal costs and revenue, we determine the x-value at which the two lines in the graph intersect. The fixed cost of $30000 represents the y-intercept of the cost line, and the variable cost is $5 per box. Therefore, the equation of the cost line is: $y = 5x + 30000$. The revenue line represents the income from selling x boxes of cookies, at \$15 per box. The equation of the revenue line is: $y = 15x$. Setting the two equations equal to each other: $5x + 30000 = 15x$. Subtracting $5x$ from both sides gives: $30000 = 10x$. Dividing both sides by 10 gives: $x = 3000$. Therefore, the bakery needs to sell 3000 boxes of cookies to achieve equal costs and revenue.

46) The x-intercepts of a parabola $h(x) = x^2 - 10x + 21$ are the solutions to the equation $h(x) = 0$. Solving $0 = x^2 - 10x + 21$ leads to the quadratic equation $(x-3)(x-7) = 0$, giving x-intercepts at $x = 3$ and $x = 7$. The distance between these intercepts is $7 - 3 = 4$ units.

47) If $x - a$ is a factor of $p(x)$, then $p(a) = 0$. The table shows that $p(-2) = 0$, suggesting that $x + 2$ is a factor of $p(x)$.

48) The solution of the equation $y = 6$ is equivalent to the intersection of the graph $y = 2x - 4$ and the horizontal line $y = 6$. Observe the graph below. The intersection is the ordered pair $(5,6)$. The first component of this ordered pair is 5, which is the solution to the problem. Therefore, choice D is the correct answer.

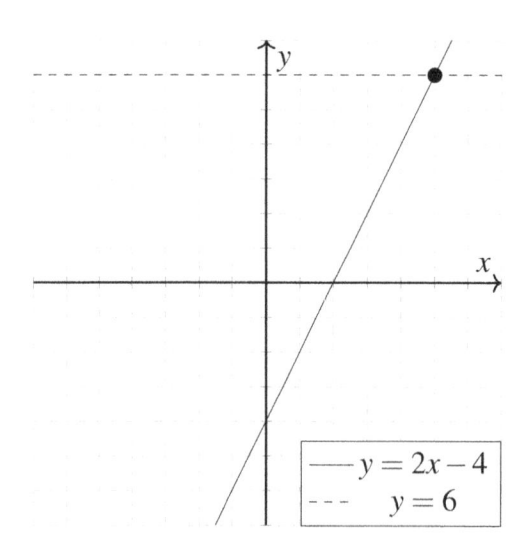

49) Let h represent the number of hardcover books sold. The number of paperback books sold is then $200 - h$. The revenue from hardcover books is $15h$, and from paperback books is $10(200 - h)$. The total revenue is between \$2500 and \$3000, so $2500 < 15h + 10(200 - h) < 3000$. Simplifying this inequality gives $2500 < 5h + 2000 < 3000$. Subtracting 2000 from all parts of the inequality yields $500 < 5h < 1000$, and dividing by 5 gives $100 < h < 200$. Therefore, the possible range for h is $100 < h < 200$.

50) For any given y value, the x values on the graph of k are half of what they would be on the graph of h.

5.3 Answers with Explanation

This makes the graph of k narrower than the graph of h.

6. Practice Test 5

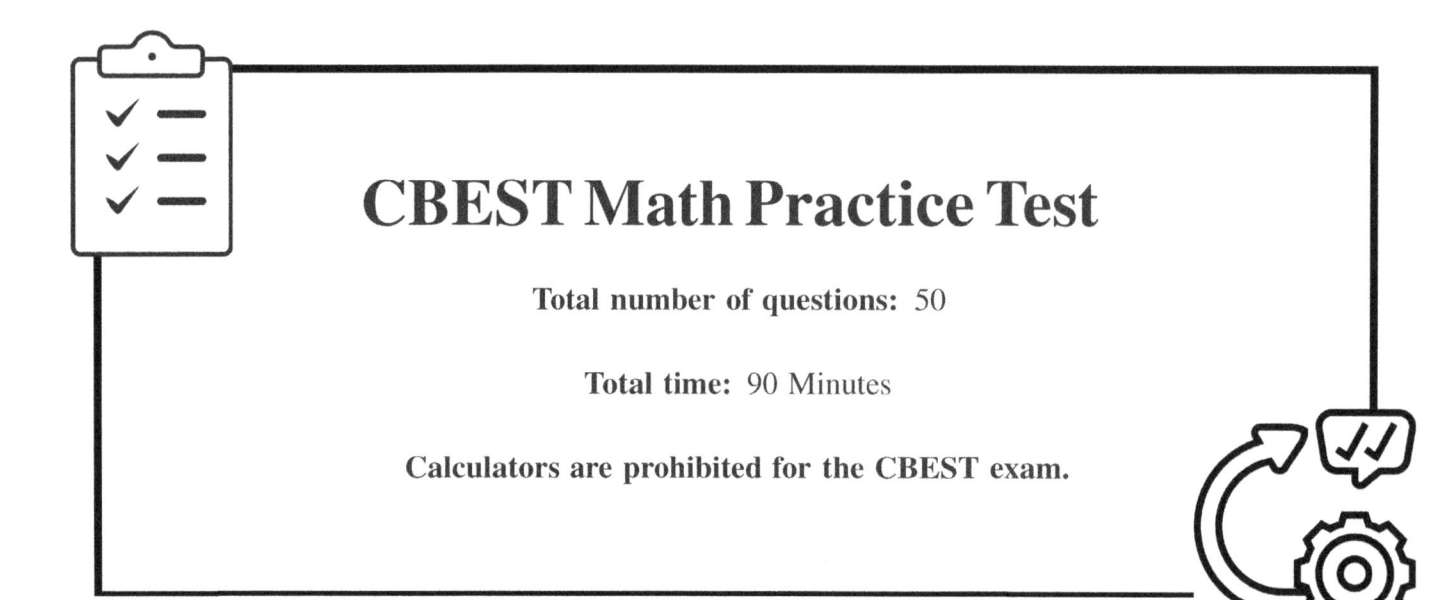

CBEST Math Practice Test

Total number of questions: 50

Total time: 90 Minutes

Calculators are prohibited for the CBEST exam.

6.1 Practices

1) A loan with a 4% simple annual interest rate is taken out for $7000. No additional amounts are borrowed. What total interest amount will be due at the end of 3 years?

☐ A. $840
☐ B. $880
☐ C. $910
☐ D. $950

6.1 Practices

☐ E. $1000

2) Which of the following situations does NOT represent a proportional relationship?

☐ A. The amount of paint needed to cover x square feet in a room, with 3 gallons of paint needed for every 100 square feet.

☐ B. The total weight w of x bags of groceries, with each bag weighing 5 pounds.

☐ C. The cost c for x hours of parking at a rate of $4 per hour.

☐ D. The length of fabric f in yards required for x dresses, with each dress requiring 2.5 yards of fabric.

☐ E. The amount of money m earned for x hours of work at a varying rate of $10 for the first hour and $15 for each subsequent hour.

3) The circumference of a circle is equal to the perimeter of a rectangle. The radius of the circle is 3 units. If the length of the rectangle is twice its width, what is the width of the rectangle? (Use $\pi = 3$)

☐ A. 1.5 units

☐ B. 2 units

☐ C. 3 units

☐ D. 4.5 units

☐ E. 6 units

4) Quadrilateral $WXYZ$ is graphed on a coordinate grid with vertices at $W(-4,3)$, $X(-1,5)$, $Y(2,1)$, and $Z(-2,-3)$. Quadrilateral $WXYZ$ is rotated 90° clockwise around the origin to create quadrilateral $W'X'Y'Z'$. Which ordered pair represents the coordinates of the vertex X'?

☐ A. $(-5,-1)$

☐ B. $(3,4)$

☐ C. $(-3,-4)$

☐ D. $(5,1)$

☐ E. $(-1,5)$

5) A local bakery tracks the number of customers per day and the total sales in dollars for that day. The data is represented in the following table. A linear function can model this data.

Number of Customers	Total Sales ($)
50	200
80	320
100	400
120	480
140	560

Based on the table, what is the best prediction for the total sales if the bakery has 150 customers in a day?

- ☐ A. $450
- ☐ B. $500
- ☐ C. $550
- ☐ D. $600
- ☐ E. $650

6) Two friends are saving money for a trip. One friend saves $5 per day and has already saved $100. The other friend saves $7 per day and has already saved $50. Which equation can be used to find d, the number of days they each need to save so that the total amount saved is the same for both friends?

- ☐ A. $5d + 50 = -7d + 100$
- ☐ B. $5d + 50 = 7d + 100$
- ☐ C. $5d + 100 = 7d + 50$
- ☐ D. $-7d + 50 = 5d + 100$
- ☐ E. $7d + 100 = 5d - 50$

7) Ms. Garcia opens a savings account with $3000.

- The bank offers 3.5% interest compounded annually on this account.
- Ms. Garcia makes no additional deposits or withdrawals.

Which amount is closest to the balance of the account at the end of 4 years?

- ☐ A. $3412.55
- ☐ B. $3450.98
- ☐ C. $3442.57
- ☐ D. $3576.89

6.1 Practices

☐ E. $3600.75

8) Point $C(-3,-4)$ and Point $D(2,6)$ are located on a coordinate grid. What is the distance between Point C and Point D in units?

☐ A. 5 units

☐ B. $\sqrt{120}$ units

☐ C. 10 units

☐ D. $\sqrt{125}$ units

☐ E. 12 units

9) A parallelogram has a perimeter of 24 units and its area is three times the perimeter. If the base of the parallelogram is $\frac{1}{3}$ of the height, what is the height of the parallelogram in units? (Choose the closest value)

☐ A. 6 units

☐ B. 9 units

☐ C. 15 units

☐ D. 18 units

☐ E. 24 units

10) A system of linear equations is represented by line h and line k. Below is a table representing some points on line h and a description for the graph of line k.

Table for Line h:

x	y
0	3
1	2
2	1

Description for Line k: Line k passes through the points $(0,4)$ and $(2,0)$.

Which system of equations is best represented by lines h and k?

☐ A. $\begin{cases} y = -x+5 \\ y = \frac{1}{2}x - 1 \end{cases}$

☐ B. $\begin{cases} 3x + 2y = 6 \\ x - 2y = 4 \end{cases}$

☐ C. $\begin{cases} x+y=3 \\ 2x+y=4 \end{cases}$

☐ D. $\begin{cases} y=2x+1 \\ y=-\frac{3}{4}x+3 \end{cases}$

☐ E. $\begin{cases} x+3y=7 \\ 2x-y=1 \end{cases}$

11) A cylindrical container has a diameter of 8 inches and a height of 15 inches. What is the volume of the container in cubic inches? (Use $\pi = 3$)

☐ A. 360 in^3

☐ B. 480 in^3

☐ C. 600 in^3

☐ D. 720 in^3

☐ E. 960 in^3

12) What is the median of these numbers? 3, 11, 7, 14, 9, 16, 6

☐ A. 7

☐ B. 8

☐ C. 9

☐ D. 11

☐ E. 14

13) Simplify $\frac{8}{\sqrt{18}-4}$.

☐ A. 2

☐ B. $\sqrt{18}$

☐ C. $\sqrt{18}+4$

☐ D. $4\sqrt{18}$

☐ E. $4(\sqrt{18}+4)$

14) What is the value of the y-intercept of the graph of $f(x) = 15(1.5)^{x-2}$?

☐ A. 6.67

6.1 Practices

- ☐ B. 10.67
- ☐ C. 15.34
- ☐ D. 22.5
- ☐ E. 30

15) Given that Jack currently has 8 books on his shelf and adds 3 new books every month, which representation best shows the relationship between the number of books, y, on Jack's shelf and the number of months that have passed, x?

- ☐ A. $y = 3x + 8$
- ☐ B. $y = 8x + 3$
- ☐ C. $y = 3x - 8$
- ☐ D. $y = 8x - 3$
- ☐ E. $y = \frac{x}{3} + 8$

16) Mark traveled 120 km in 3 hours and Lisa traveled 200 km in 5 hours. What is the ratio of the average speed of Mark to the average speed of Lisa?

- ☐ A. $2:3$
- ☐ B. $3:2$
- ☐ C. $1:1$
- ☐ D. $5:6$
- ☐ E. $6:5$

17) Identify the set of ordered pairs that define y as a function of x.

- ☐ A. $\{(1,4),(1,2),(3,6),(4,8)\}$
- ☐ B. $\{(2,3),(5,4),(6,7),(3,2)\}$
- ☐ C. $\{(0,5),(2,6),(0,7),(3,8)\}$
- ☐ D. $\{(-2,1),(1,2),(-2,3),(4,5)\}$
- ☐ E. $\{(7,3),(8,4),(9,5),(7,6)\}$

18) For the function $f(x) = 2x^2 - 9x + 4$, what is a true statement?

- ☐ A. The zeroes are $-\frac{1}{2}$ and 4, because the factors of f are $(2x+1)$ and $(x-4)$.
- ☐ B. The zeroes are $\frac{1}{2}$ and 4, because the factors of f are $(2x-1)$ and $(x-4)$.

☐ C. The zeroes are $\frac{1}{2}$ and -4, because the factors of f are $(2x+1)$ and $(x+4)$.

☐ D. The zeroes are 2 and 4, because the factors of f are $(x-2)$ and $(x-4)$.

☐ E. The zeroes are 2 and -4, because the factors of f are $(x-2)$ and $(x+4)$.

19) An inequality is shown: $0.5 < x < 2.0$, which value of x makes the inequality true? (There may be multiple options.)

☐ A. 0.3

☐ B. 0.6

☐ C. 1.8

☐ D. 2.1

☐ E. 2.3

20) Mrs. Smith opened an account with a deposit of $5000.

- The account earned annual simple interest.

- She did not make any additional deposits or withdrawals.

- At the end of 4 years, the balance of the account was $6000.

What is the annual interest rate on this account?

☐ A. 2%

☐ B. 4%

☐ C. 5%

☐ D. 6%

☐ E. 8%

21) The graph shows the relationship between the number of pages in a notebook and the number of days since it started being used.

6.1 Practices

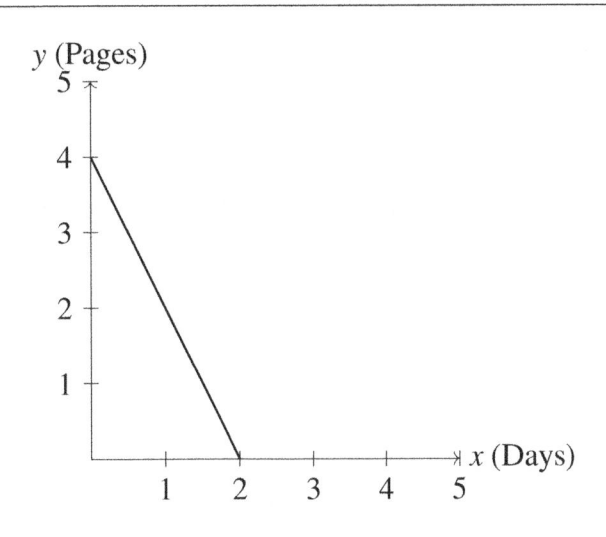

Which equation can be used to find y, which is the number of pages remaining in the notebook after x days of use?

☐ A. $y = -2x + 4$

☐ B. $y = -\frac{1}{2}x + 4$

☐ C. $y = 2x - 4$

☐ D. $y = 2x + 4$

☐ E. $y = -2x - 4$

22) The table below shows the linear relationship between the distance traveled by a taxi and the fare charged.

Distance traveled(x)	Fare charged(y)
2	8
4	14
6	20
8	26
10	32

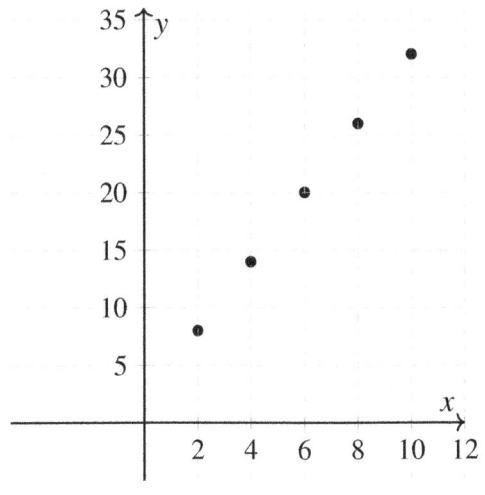

What is the slope of the line that represents this relationship?

☐ A. $2 per mile

☐ B. $3 per mile

☐ C. $4 per mile

☐ D. $5 per mile

☐ E. $6 per mile

23) A chemical solution contains 3% salt. If there are 18 ml of salt, what is the volume of the solution?

☐ A. 300 ml

☐ B. 400 ml

☐ C. 600 ml

☐ D. 800 ml

☐ E. 1000 ml

24) The value of x varies directly with y. When $x = 80$, $y = 5$. What is the value of x when y is 8?

☐ A. 25

☐ B. 64

☐ C. 128

☐ D. 160

☐ E. 320

25) Mark is saving $30 that he earned from a garage sale. He earns $15 every week for mowing lawns, which he also saves.

Which function can be used to find m, which is the amount of money Mark will have saved at the end of t weeks of mowing lawns?

☐ A. $m = 15t + 30$

☐ B. $m = 25t$

☐ C. $m = 30t + 15$

☐ D. $m = 45t$

☐ E. $m = 60t$

26) A gardener is planning to create a square flower bed in a park. If the area of the flower bed is planned to be 196 square meters, what will be the length of each side of the flower bed?

☐ A. 12m

☐ B. 13m

☐ C. 14m

6.1 Practices

☐ D. 16m

☐ E. 98m

27) If $\frac{y+4}{7} = M$ and $M = 5$, what is the value of y?

☐ A. 11

☐ B. 31

☐ C. 35

☐ D. 39

☐ E. 40

28) Emily plans to buy a laptop worth $800. She has agreed to pay 40% of the cost, while her parents will cover the rest. She wants to save up for it over the next 4 months. What is the minimum amount she needs to save each month to purchase the laptop?

☐ A. $50

☐ B. $80

☐ C. $100

☐ D. $200

☐ E. $320

29) A chef needs to prepare a large batch of a recipe that requires $\frac{2}{3}$ pound of sugar per cake. If the total amount of sugar available is 60 pounds, how many cakes can the chef make?

☐ A. 30

☐ B. 40

☐ C. 60

☐ D. 90

☐ E. 120

30) Lara has a collection of novels and short stories. She has one shelf that contains 30 novels. She has a second shelf that has 10 short stories in each row. The equation below can be used to find n, the total number of books Lara has, if the second shelf has r rows:

$$n = 30 + 10r.$$

How many books does Lara have in total if the second shelf has 8 rows?

- [ ] A. 80
- [ ] B. 110
- [ ] C. 120
- [ ] D. 140
- [ ] E. 160

31) A city health official surveyed 150 randomly selected residents about their exercise habits. Of those surveyed, 45 residents said they exercise at least three times a week. Based on these results, how many of the 10000 residents in the city are expected to exercise at least three times a week?

- [ ] A. 3000
- [ ] B. 4500
- [ ] C. 6000
- [ ] D. 7500
- [ ] E. 9000

32) Quadratic function $f(x)$ represents the trajectory, in meters, of a ball thrown upwards from a height of 5 meters. The graph of the function is shown below.

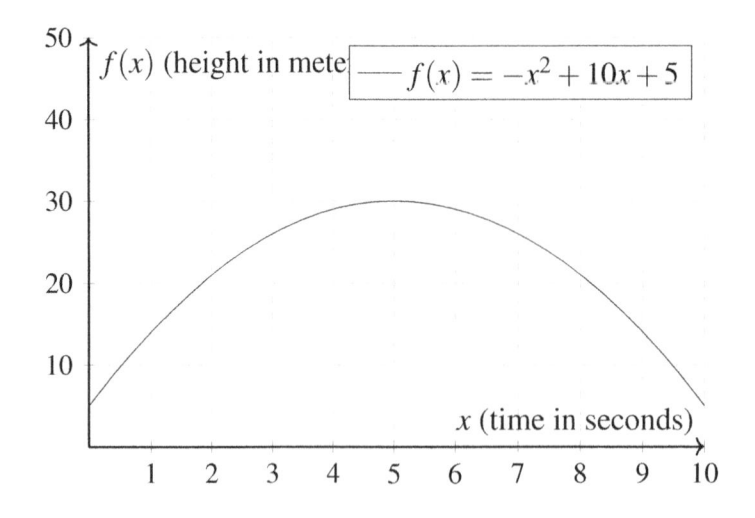

$$f(x) = -x^2 + 10x + 5$$

What is the maximum value of the graph of the function?

- [ ] A. 45
- [ ] B. 25
- [ ] C. 30

6.1 Practices

☐ D. 40

☐ E. 35

33) box contains the following items:

Number of items	Item
5	Marbles
10	Coins
3	Dice

A person will randomly select an item from the box, and then replace it. Then, they will randomly select another item from the box. What is the probability that a coin will be selected both times?

☐ A. $\frac{1}{9}$

☐ B. $\frac{36}{81}$

☐ C. $\frac{1}{4}$

☐ D. $\frac{25}{81}$

☐ E. $\frac{5}{18}$

34) A hiker travels 24 miles north and then 7 miles west. How far is the hiker from the starting point?

☐ A. 25 miles

☐ B. 26 miles

☐ C. 27 miles

☐ D. 28 miles

☐ E. 31 miles

35) A school is buying textbooks and notebooks for a class. The textbooks cost between $20 and $30 each, and the notebooks cost between $2 and $4 each. Which of these does NOT represent a reasonable total purchase price for 15 textbooks and 40 notebooks?

☐ A. $400

☐ B. $450

☐ C. $500

☐ D. $550

☐ E. $650

36) How many tiles of 12 cm² are needed to cover a tabletop of dimension 9 cm by 36 cm?

- ☐ A. 9
- ☐ B. 18
- ☐ C. 27
- ☐ D. 30
- ☐ E. 36

37) What is the domain of $g(x) = 3x^2 - 16$?

- ☐ A. $(-\infty, 16]$
- ☐ B. $(-4, 4)$
- ☐ C. $\left[-\frac{4}{3}, \frac{4}{3}\right]$
- ☐ D. $\mathbb{R}$
- ☐ E. $[-3, \infty)$

38) Which scatterplot suggests a linear relationship between x and y?

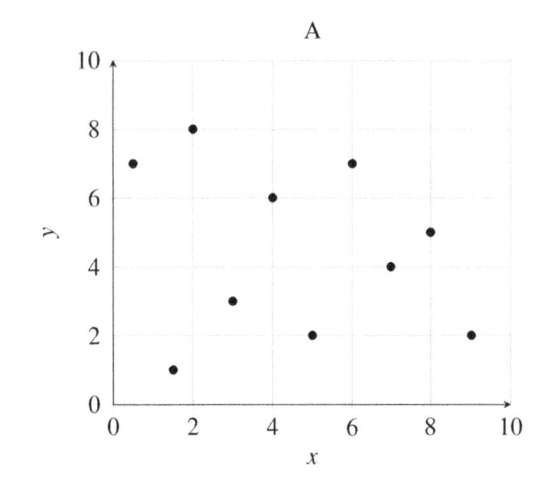

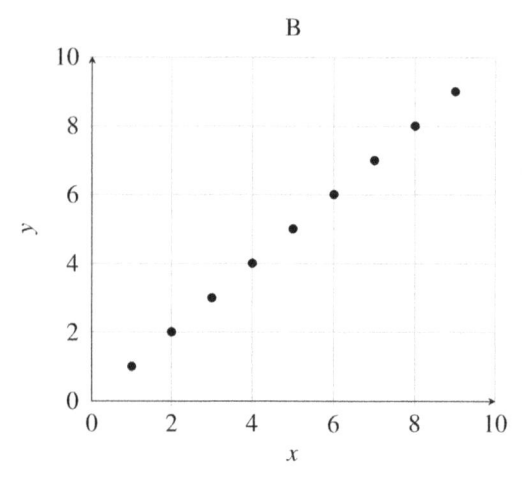

6.1 Practices

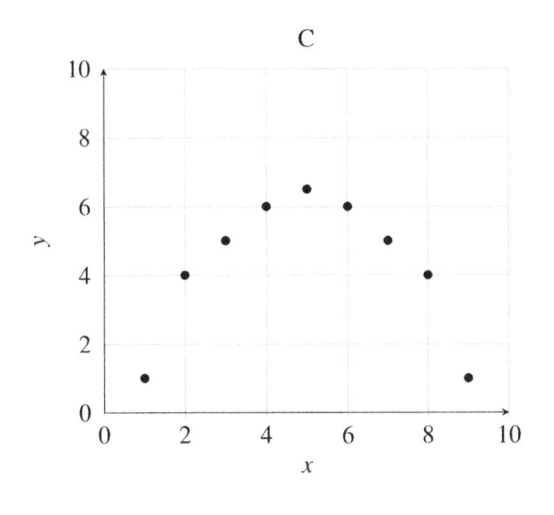

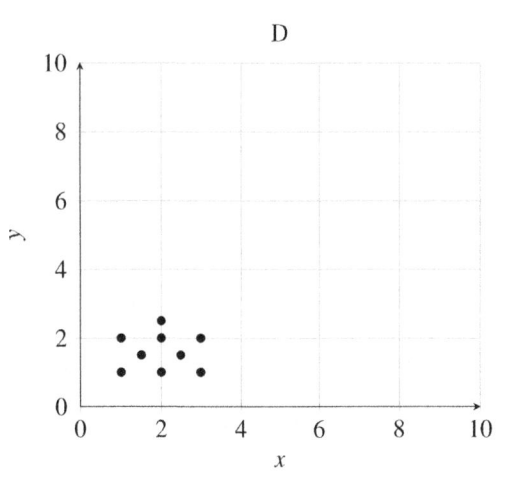

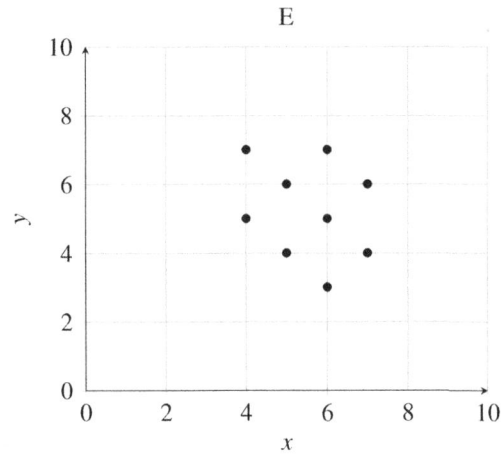

C

D

E

☐ A.

☐ B.

☐ C.

☐ D.

☐ E.

39) Jordan has $2000 to deposit into two different savings accounts.

- Jordan will deposit $1200 into Account X, which earns 5% annual simple interest.

- He will deposit $800 into Account Y, which earns 4.5% interest compounded annually.

- Jordan will not make any additional deposits or withdrawals.

Which amount is closest to the total balance of these two accounts at the end of 4 years? (Choose the closest option)

☐ A. $600

☐ B. $1080

☐ C. $2000

☐ D. $2280

☐ E. $2394

40) Suppose a recipe calls for $\frac{1}{4}$ cup of flour, $\frac{1}{3}$ tablespoon of baking powder, and 20% of a cup of milk. What is the total amount of these ingredients, listed from greatest to least?

☐ A. $\frac{1}{4}$ cup, 20% cup, $\frac{1}{3}$ tablespoon

☐ B. 20% cup, $\frac{1}{3}$ tablespoon, $\frac{1}{4}$ cup

☐ C. 20% cup, $\frac{1}{4}$ cup, $\frac{1}{3}$ tablespoon

☐ D. $\frac{1}{3}$ tablespoon, 20% cup, $\frac{1}{4}$ cup

☐ E. $\frac{1}{4}$ cup, $\frac{1}{3}$ tablespoon, 20% cup

41) Three fourths of 24 is equal to $\frac{3}{8}$ of what number?

☐ A. 16

☐ B. 24

☐ C. 32

☐ D. 48

☐ E. 64

42) Which representation of a transformation on a coordinate grid does not preserve congruence?

☐ A. $(x,y) \to (x-3, y-3)$

☐ B. $(x,y) \to (x,-y)$

☐ C. $(x,y) \to (\frac{x}{2}, \frac{y}{2})$

☐ D. $(x,y) \to (y,-x)$

☐ E. $(x,y) \to (-y,x)$

43) Which of the following is equivalent to $3\sqrt{18} + 3\sqrt{2}$?

☐ A. $\sqrt{2}$

☐ B. 3

☐ C. $6\sqrt{2}$

☐ D. $9\sqrt{2}$

6.1 Practices

☐ E. $12\sqrt{2}$

44) The graph of a quadratic function is shown on the grid.

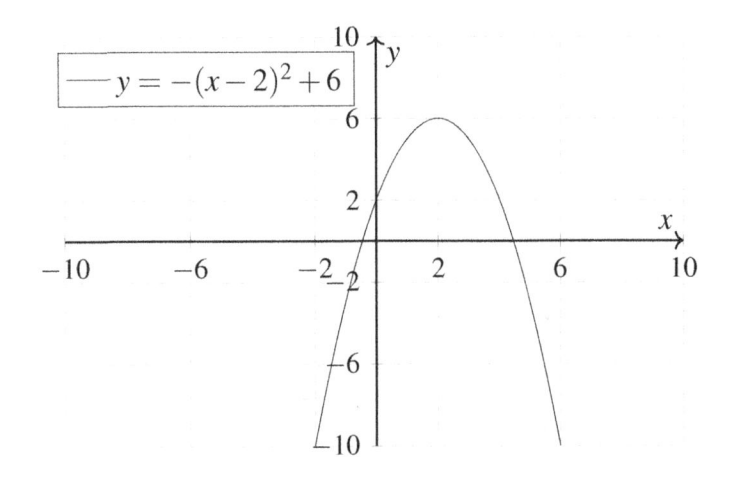

$y = -(x-2)^2 + 6$

Which equation best represents the axis of symmetry?

☐ A. $x = -3$

☐ B. $x = 2$

☐ C. $x = 4$

☐ D. $y = 0$

☐ E. $y = 6$

45) An editor earns $400 per week for working 30 hours plus $15 per hour for any hours worked over 30 hours. She can work a maximum of 54 hours per week.

Which graph best represents the editor's weekly earnings in dollars for working h hours in a week?

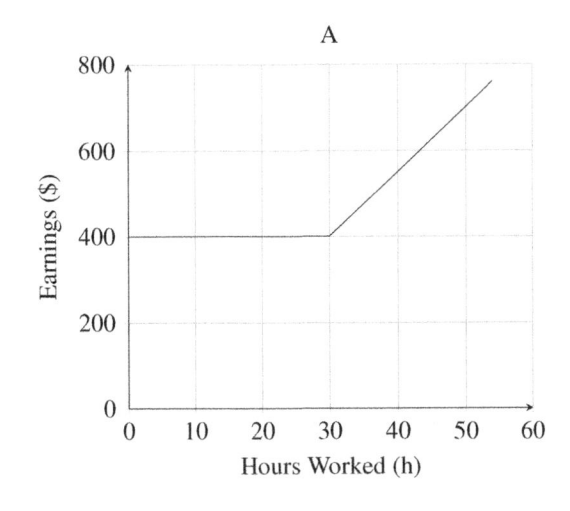

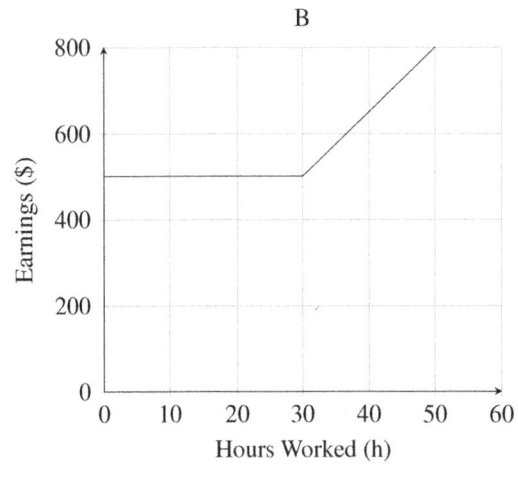

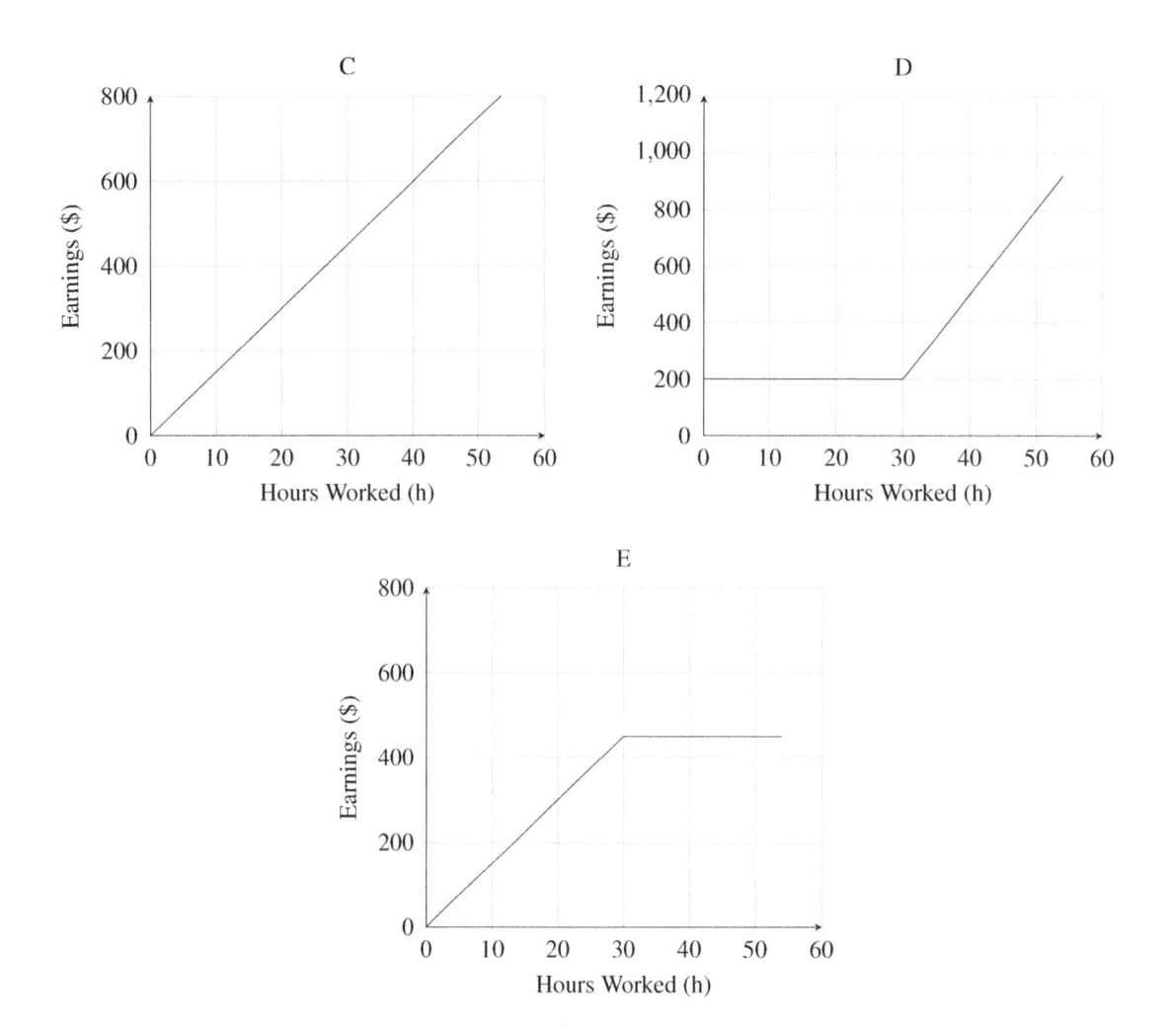

□ A. A graph with a constant line at $400 from 0 to 30 hours, then a linearly increasing line from 30 to 54 hours.

□ B. A graph with a constant line at $415 from 0 to 30 hours, then a linearly increasing line from 30 to 54 hours.

□ C. A linearly increasing line from 0 to 54 hours.

□ D. A graph with a constant line at $400 from 0 to 30 hours, then a steeply increasing line from 30 to 54 hours.

□ E. A graph that increases linearly from 0 to 30 hours, then remains constant from 30 to 54 hours.

46) The conversion from feet to inches can be represented by a linear relationship. The graph shows the linear relationship between y, the length in inches, and x, the length in feet. Assume that 1 foot is approximately 12 inches.

6.1 Practices

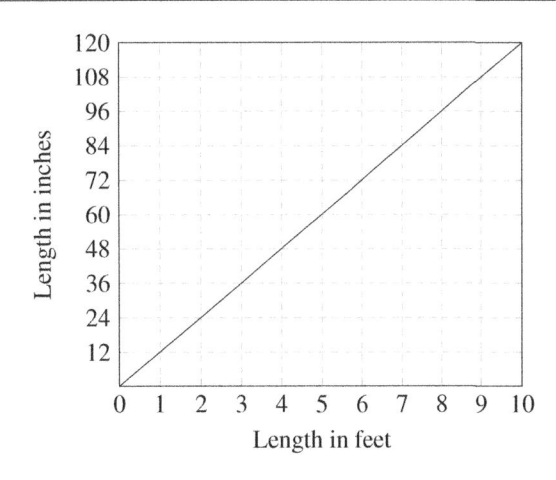

Which equation best represents this situation?

- ☐ A. $y = 12x$
- ☐ B. $y = 12(x-1)$
- ☐ C. $y = 12(x+1)$
- ☐ D. $y = \frac{1}{12}(x-1)$
- ☐ E. $y = \frac{1}{12}x + 1$

47) Which of the following is the same as: $0.000,000,000,005,678$?

- ☐ A. 5.678×10^{-12}
- ☐ B. 5.678×10^{-10}
- ☐ C. 56.78×10^{-12}
- ☐ D. 56.78×10^{-13}
- ☐ E. 5.678×10^{11}

48) Create a cubic function that has roots at -2, 1, and 3.

- ☐ A. $f(x) = (x+2)(x-1)(x+3)$
- ☐ B. $f(x) = x^3 - 2x^2 - x + 6$
- ☐ C. $f(x) = x^3 - 2x^2 - 5x + 6$
- ☐ D. $f(x) = x^3 + x^2 - 6x$
- ☐ E. $f(x) = x^3 - 2x^2 + x - 3$

49) Simplify the expression $(2a^2b)^3(4ab^2)^2$.

- ☐ A. $128a^8b^7$

- [ ] B. $182a^5b^5$
- [ ] C. $16a^6b^6$
- [ ] D. $32a^7b^7$
- [ ] E. $256a^5b^7$

50) The graph of the linear function g is shown on the grid.

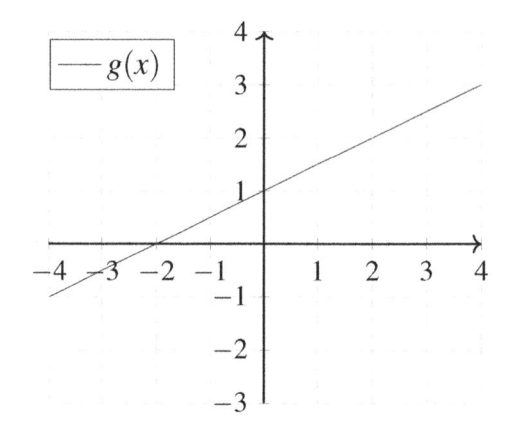

What is the zero of g?

- [ ] A. -3
- [ ] B. -2
- [ ] C. 0
- [ ] D. 2
- [ ] E. 3

6.2 Answer Keys

1) A. $840
2) E.
3) C. 3 units
4) D. (5,1)
5) D. $600
6) C. $5d + 100 = 7d + 50$
7) C. $3,442.57
8) D. $\sqrt{125}$ units
9) C. 15 units
10) C. $\begin{cases} x+y=3 \\ 2x+y=4 \end{cases}$
11) D. $720\,\text{in}^3$
12) C. 9
13) E. $4(\sqrt{18}+4)$
14) A. 6.67
15) A. $y = 3x + 8$
16) C. 1 : 1
17) B. $\{(2,3),(5,4),(6,7),(3,2)\}$
18) B.
19) B. 0.6 and C. 1.8
20) C. 5%
21) A. $y = -2x + 4$
22) B. $3 per mile
23) C. 600 ml
24) C. 128
25) A. $m = 15t + 30$
26) C. 14m
27) B. 31
28) B. $80
29) D. 90
30) B. 110
31) A. 3000
32) C. 30
33) D. $\frac{25}{81}$
34) A. 25 miles
35) E. $650
36) C. 27
37) D. $\mathbb{R}$
38) B.
39) E. $2,394
40) A. $\frac{1}{4}$ cup, 20% cup, $\frac{1}{3}$ tablespoon
41) D. 48
42) C. $(x,y) \to (\frac{x}{2}, \frac{y}{2})$
43) E. $12\sqrt{2}$
44) B. $x = 2$
45) A.
46) A. $y = 12x$
47) A. 5.678×10^{-12}
48) C. $f(x) = x^3 - 2x^2 - 5x + 6$
49) A. $128a^8 b^7$
50) B. -2

6.3 Answers with Explanation

1) To calculate simple interest, use the formula $I = P \times r \times t$, where I is the interest, P is the principal amount, r is the rate of interest, and t is the time in years. Here, $P = \$7000$, $r = 4\% = 0.04$, and $t = 3$ years. So, $I = \$7000 \times 0.04 \times 3 = \$840$. Hence, the total interest amount at the end of 3 years is $\$840$.

2) A proportional relationship is one where the ratio between two quantities remains constant.

In options A, B, C, and D, the ratios (paint per square feet, weight per bag, cost per hour, and fabric per dress) are constant, representing proportional relationships.

However, in option E, the rate of earning changes after the first hour (from $10 to $15 per hour). This means the ratio of money earned to hours worked is not constant, and thus, it does not represent a proportional relationship.

3) The circumference of a circle is given by $C = 2\pi r$. For a circle with radius 3 units and using $\pi = 3$, $C = 2 \times 3 \times 3 = 18$ units.

For the rectangle, let the width be w and the length be $2w$. The perimeter of the rectangle is $P = 2(l+w) = 2(2w+w) = 6w$.

Since the circumference of the circle is equal to the perimeter of the rectangle, $18 = 6w$. Solving for w, we get $w = \frac{18}{6} = 3$ units.

Therefore, the width of the rectangle is 3 units.

4) When a point (x,y) is rotated $90°$ clockwise around the origin, the new coordinates become $(y, -x)$.

For vertex $X(-1, 5)$, applying this transformation, the coordinates of X' become $(5, -(-1)) = (5, 1)$. Thus, the coordinates of vertex X' after the rotation are $(5, 1)$.

5) To predict the total sales based on the number of customers, first determine the linear relationship from the provided data in the table. The table suggests that for every increase of 30 customers, the sales increase by $120.

This indicates that for each additional customer, the sales increase by $4 ($120 divided by 30 customers). Therefore, for 150 customers, the total sales would be $150 \times \$4 = \$600$. Thus, the best prediction for the total sales is $600.

6.3 Answers with Explanation

6) To find the number of days d for which both friends have saved the same amount, set up an equation that equates their total savings.

Friend 1 saves $5 per day and has already saved $100. Therefore, their total savings after d days will be $5d + 100$. Friend 2 saves $7 per day and has already saved $50. Thus, their total savings after d days will be $7d + 50$.

Equating these two expressions gives the equation $5d + 100 = 7d + 50$. Therefore, option C is correct.

7) The formula for compound interest is $A = P(1 + \frac{r}{n})^{nt}$, where:
- A is the amount of money accumulated after n years, including interest.
- P is the principal amount (the initial amount of money).
- r is the annual interest rate (as a decimal).
- n is the number of times that interest is compounded per year.
- t is the time the money is invested for in years.

Given $P = \$3000$, $r = 3.5\% = 0.035$, $n = 1$ (compounded annually), and $t = 4$ years, the calculation is:

$$A = 3000 \times \left(1 + \frac{0.035}{1}\right)^{(1 \times 4)} = \$3442.57.$$

Therefore, the balance at the end of 4 years is $3442.57.

8) The distance between two points (x_1, y_1) and (x_2, y_2) on a coordinate grid is given by the formula

$$d = \sqrt{(x_2 - x_1)^2 + (y_2 - y_1)^2}.$$

For points $C(-3, -4)$ and $D(2, 6)$, we calculate the distance as

$$d = \sqrt{(2 - (-3))^2 + (6 - (-4))^2} = \sqrt{5^2 + 10^2} = \sqrt{25 + 100} = \sqrt{125}.$$

9) Let the height be h units and the base be $\frac{1}{3}h$ units. The area of a parallelogram is given by $A = \text{base} \times \text{height}$. The problem states that the area is three times the perimeter, so $A = 3 \times 24 = 72$ square units.

Using the area formula: $A = \frac{1}{3}h \times h = \frac{1}{3}h^2$. Setting this equal to 72, we get $\frac{1}{3}h^2 = 72$. Solving for h, we find $h^2 = 216$, and thus $h = \sqrt{216} \approx 14.7$ units. Therefore, the closest value to the height of the parallelogram among the options is 15 units.

Chapter 6. Practice Test 5

10) Based on the points from the table for line h and the slope-intercept form of line k shown in the description, the system of equations that best represents these lines is:

Line h: $x + y = 3$, which can be derived from the points on line h. Line k: $y = -2x + 4$, which can be inferred from the description of line k.

Thus, option C correctly represents the system of equations for lines h and k.

11) The volume of a cylinder is calculated using the formula $V = \pi r^2 h$, where V is the volume, r is the radius, and h is the height. The radius is half the diameter, so for a diameter of 8 inches, the radius is 4 inches.

Therefore, the volume of this cylindrical container is $V = 3 \times 4^2 \times 15 = 3 \times 16 \times 15 = 720 \, \text{in}^3$. Thus, the volume of the container is 720 cubic inches.

12) To find the median, arrange the numbers in ascending order: 3, 6, 7, 9, 11, 14, 16. Since there are 7 numbers, the median is the middle number, which is the fourth number in the sorted list. Therefore, the median of these numbers is 9.

13) To simplify $\frac{8}{\sqrt{18}-4}$, multiply the numerator and the denominator by the conjugate of the denominator, which is $\sqrt{18}+4$:

$$\frac{8}{\sqrt{18}-4} \times \frac{\sqrt{18}+4}{\sqrt{18}+4} = \frac{8(\sqrt{18}+4)}{18-16} = \frac{8(\sqrt{18}+4)}{2} = 4(\sqrt{18}+4).$$

14) The y-intercept of a graph is found by setting $x = 0$. For the function $f(x) = 15(1.5)^{x-2}$, substitute $x = 0$:

$$f(0) = 15(1.5)^{0-2} = 15(1.5)^{-2} = 15 \times \frac{1}{(1.5)^2} = 15 \times \frac{1}{2.25} = \frac{15}{2.25} = 6.67.$$

15) The relationship is a linear one where the number of books increases by 3 each month. The equation for such a relationship is of the form $y = mx + b$, where m is the rate of increase and b is the starting value. Here, Jack starts with 8 books and adds 3 books every month. Therefore, the equation representing this situation is $y = 3x + 8$. Option A, $y = 3x + 8$, correctly represents this relationship.

16) The average speed is calculated as the total distance traveled divided by the total time taken.
Mark's average speed = $\frac{120 \text{ km}}{3 \text{ hours}} = 40$ km/h.
Lisa's average speed = $\frac{200 \text{ km}}{5 \text{ hours}} = 40$ km/h.

6.3 Answers with Explanation

The ratio of Mark's speed to Lisa's speed is 40 : 40 or simplified to 1 : 1.

17) A set of ordered pairs represents a function when each input (or x-value) is associated with exactly one output (or y-value).

- In set A, the input 1 corresponds to two different outputs (4 and 2), so it is not a function.
- Set B has distinct x-values for each ordered pair, making it a function.
- In set C, the input 0 corresponds to two different outputs (5 and 7), so it is not a function.
- In set D, the input -2 corresponds to two different outputs (1 and 3), so it is not a function.
- In set E, the input 7 corresponds to two different outputs (3 and 6), so it is not a function.

Therefore, the set that represents y as a function of x is B. $\{(2,3),(5,4),(6,7),(3,2)\}$.

18) To find the zeroes of the quadratic function $f(x) = 2x^2 - 9x + 4$, we can factorize it:

$$f(x) = 2x^2 - 9x + 4 = (2x-1)(x-4).$$

Setting each factor equal to zero gives the zeroes:

$2x - 1 = 0$ gives $x = \frac{1}{2}$.

$x - 4 = 0$ gives $x = 4$.

Therefore, the correct statement is B. The zeroes are $\frac{1}{2}$ and 4, because the factors of f are $(2x-1)$ and $(x-4)$.

19) To find which value of x makes the inequality $0.5 < x < 2.0$ true, check each option:

- A. 0.3 is less than 0.5, so it does not satisfy the inequality.
- B. 0.6 is greater than 0.5 but less than 2.0, so it satisfies the inequality.
- C. 1.8 is greater than 0.5 and less than 2.0, so it satisfies the inequality.
- D. 2.1 is greater than 2.0, so it does not satisfy the inequality.
- E. 2.3 is greater than 2.0, so it does not satisfy the inequality.

Therefore, the values of x that make the inequality true are B and C.

20) The formula for simple interest is $I = P \times r \times t$, where I is the interest, P is the principal amount, r is the rate of interest, and t is the time in years. Here, $P = \$5000$, the total amount after 4 years is $\$6000$, so $I = \$6000 - \$5000 = \$1000$, and $t = 4$ years. Plugging these values into the formula, we get $1000 = 5000 \times r \times 4$. Solving for r, we find $r = \frac{1000}{5000 \times 4} = \frac{1}{20} = 0.05$, or 5%.

21) The slope of the line is calculated as $m = \frac{0-4}{2-0} = -2$. The y-intercept is 4, as the line crosses the y-axis at $y = 4$. Therefore, the equation of the line is $y = -2x + 4$, matching option A.

22) To find the slope, use the formula: $m = \frac{y_2 - y_1}{x_2 - x_1}$. Choose two points, for instance, $(2,8)$ and $(10,32)$. The slope is calculated as $\frac{32-8}{10-2} = \frac{24}{8} = 3$. Therefore, the slope of the line representing the relationship between distance and fare is $3 per mile.

23) The volume of the solution can be found using the formula for concentration:

$$\text{Concentration} = \frac{\text{Volume of Solution}}{\text{Total Volume}}.$$

Here, the concentration is 3% (or 0.03 as a decimal), and the volume of the solution (salt) is 18 ml. Let the total volume of the solution be V ml. Therefore, $0.03 = \frac{18}{V}$. Solving for V, we have $V = \frac{18}{0.03} = 600$ ml.

24) Since x varies directly with y, they are related by the equation $x = ky$, where k is the constant of variation. Given $x = 80$ when $y = 5$, we find k by plugging these values into the equation: $80 = k \times 5$ or $k = 16$. To find x when $y = 8$, substitute k and y into the equation: $x = 16 \times 8 = 128$. Therefore, the value of x when y is 8 is 128.

25) Mark starts with an initial savings of $30 from the garage sale. Each week he adds $15 to his savings from mowing lawns. The total amount saved, m, after t weeks can be represented by a linear function:

initial amount + (weekly earning × number of weeks).

Therefore, the function is $m = 15t + 30$. Option A, $m = 15t + 30$, correctly represents the relationship between the number of weeks and the total amount saved.

26) The area of a square is calculated by squaring the length of one of its sides. Let s represent the side length of the square. Given that the area is 196 square meters, we set up the equation: $s^2 = 196$. Solving for s, we find that $s = \sqrt{196} = 14$. Therefore, each side of the flower bed is 14 meters long.

27) To find the value of y, we can set up the equation: $\frac{y+4}{7} = 5$. Multiplying both sides by 7 gives: $y + 4 = 35$. Subtracting 4 from both sides yields: $y = 35 - 4 = 31$. Therefore, the value of y is 31.

28) First, calculate the amount Emily has to pay: 40% of $800 is $800 × 0.40 = $320. To find out how much she needs to save each month over 4 months, divide the total amount by 4: $320 \div 4 = \$80$. Therefore, Emily needs to save $80 each month.

29) To find out how many cakes can be made with 60 pounds of sugar, when each cake requires $\frac{2}{3}$ pound of sugar, divide the total amount of sugar by the amount needed per cake:

$$60 \div \frac{2}{3} = 60 \times \frac{3}{2} = 90.$$

Therefore, the chef can make 90 cakes.

30) Substitute $r = 8$ into the equation:

$$n = 30 + (10 \times 8) = 30 + 80 = 110.$$

Therefore, Lara has a total of 110 books.

31) To estimate the number of residents who exercise at least three times a week, set up a proportion:

$$\frac{45 \text{ exercise}}{150 \text{ surveyed}} = \frac{x \text{ exercise}}{10,000 \text{ total residents}}.$$

Solving for x, we get

$$x = \frac{45}{150} \times 10000 = 3000.$$

Therefore, it is estimated that 3000 residents exercise at least three times a week.

32) The maximum value of a quadratic function $f(x) = ax^2 + bx + c$ can be found by calculating the vertex of the parabola. The x-coordinate of the vertex is given by $-\frac{b}{2a}$. In this case, $a = -1$ and $b = 10$, so the x-coordinate is $-\frac{10}{2 \times (-1)} = 5$. Substituting $x = 5$ into the function gives the maximum height:

$$f(5) = -(5)^2 + (10 \times 5) + 5 = -25 + 50 + 5 = 30.$$

Therefore, the maximum value of the graph of the function is 30 meters.

33) The total number of items in the box is $5 + 10 + 3 = 18$. The probability of selecting a coin on the first draw is $\frac{10}{18}$. Since the coin is replaced, the probability of selecting a coin on the second draw is also $\frac{10}{18}$. The probability of both events happening is the product of the individual probabilities:

$$\frac{10}{18} \times \frac{10}{18} = \frac{100}{324} = \frac{25}{81}.$$

Therefore, the probability that a coin will be selected both times is $\frac{25}{81}$.

34) The distance from the starting point can be found using the Pythagorean theorem. Let the northward travel be one leg of a right triangle, and the westward travel be the other leg. The distance from the starting point is the hypotenuse. Therefore, the calculation is:

$$\sqrt{(24 \text{ miles})^2 + (7 \text{ miles})^2} = \sqrt{576 + 49} = \sqrt{625} = 25 \text{ miles}.$$

Hence, the hiker is 25 miles from the starting point.

35) To find the reasonable range for the total cost, calculate the minimum and maximum possible costs. For textbooks, the range is $\$20 \times 15 = \$300$ to $\$30 \times 15 = \$450$. For notebooks, the range is $\$2 \times 40 = \$80$ to $\$4 \times 40 = \$160$. The total range is from $\$380$ to $\$610$. Therefore, a total of $\$650$ is not a reasonable purchase price as it exceeds the maximum possible cost.

36) First, calculate the area of the tabletop: $9 \text{ cm} \times 36 \text{ cm} = 324 \text{ cm}^2$. Then, divide this area by the area of one tile to find the number of tiles needed:

$$\frac{324 \text{ cm}^2}{12 \text{ cm}^2} = 27.$$

Therefore, 27 tiles are needed to cover the tabletop.

37) The function $g(x) = 3x^2 - 16$ is a quadratic function. Quadratic functions are defined for all real numbers. Therefore, the domain of $g(x)$ is all real numbers, denoted as $\mathbb{R}$.

38) A linear relationship between two variables is represented by points forming a straight line in a scatterplot. Option B, which describes a scatterplot with points forming a perfect straight line, is indicative of a linear relationship between x and y.

39) For Account X: Simple interest is calculated as $\$1200 \times 5\% \times 4 = \$240$. The total balance in Account X

6.3 Answers with Explanation 149

after 4 years is $1200 + $240 = $1440.

For Account Y: Compounded annually, the formula is $P(1+r/n)^{nt}$. Here, $P = \$800$, $r = 4.5\%$, $n = 1$, and $t = 4$. The total balance in Account Y after 4 years is $\$800 \times (1+0.045)^4 \approx \$954.01$.

Adding the totals from both accounts gives approximately $\$1440 + \$954.01 = \$2394.01$. So, the closest option is E.

40) First, convert the percentage to a fraction: 20% of a cup is $\frac{1}{5}$ cup. Now, compare the quantities: $\frac{1}{4}$ cup (flour) is greater than $\frac{1}{5}$ cup (milk), which is greater than $\frac{1}{3}$ tablespoon (baking powder). Thus, the order from greatest to least is $\frac{1}{4}$ cup, 20% cup, $\frac{1}{3}$ tablespoon.

41) Let the unknown number be x. The equation is:

$$\frac{3}{4} \times 24 = \frac{3}{8} \times x.$$

Solving for x, we get:

$$18 = \frac{3}{8}x \Rightarrow x = \frac{18 \times 8}{3} = 48.$$

Therefore, three fourths of 24 is equal to $\frac{3}{8}$ of 48.

42) Transformations that preserve congruence are isometries, which include translations, reflections, and rotations. These transformations do not change the size of the figure. The transformation $(x,y) \rightarrow (\frac{x}{2}, \frac{y}{2})$ is a scaling transformation that reduces the size of the figure, thus not preserving congruence.

43) Simplify $\sqrt{18}$ as $\sqrt{9 \times 2} = 3\sqrt{2}$. Thus, the expression becomes:

$$3 \times 3\sqrt{2} + 3\sqrt{2} = 9\sqrt{2} + 3\sqrt{2} = 12\sqrt{2}.$$

Therefore, the equivalent expression is $12\sqrt{2}$.

44) The axis of symmetry of a quadratic function $y = a(x-h)^2 + k$ is the vertical line $x = h$. For the given function $y = -(x-2)^2 + 6$, the axis of symmetry is at $x = 2$.

45) For the first 30 hours, the editor earns a flat rate of $400, represented by a horizontal line. Beyond 30 hours, she earns an additional $15 per hour, so the graph should show a linear increase from 30 to 54 hours.

46) Observing the graph, we see that the line passes through the points $(0,0)$ and $(10,120)$. Therefore, the slope m can be calculated as:
$$m = \frac{y_2 - y_1}{x_2 - x_1} = \frac{120-0}{10-0} = \frac{120}{10} = 12.$$

The y-intercept is the point where the line crosses the y-axis. In this graph, the line crosses the y-axis at $(0,0)$. This means the y-intercept is 0.

Combining the slope and y-intercept, the equation of the line is $y = mx + b$, where m is the slope and b is the y-intercept. Substituting in our values, we get: $y = 12x + 0$, which simplifies to: $y = 12x$. Therefore, based solely on the graph, the equation $y = 12x$ best represents the linear relationship between the length in feet (x) and the length in inches (y).

47) To convert the number $0.000,000,000,005,678$ to scientific notation, move the decimal point 12 places to the right, which gives 5.678×10^{-12}. Therefore, the equivalent expression in scientific notation is 5.678×10^{-12}.

48) A cubic function with roots at -2, 1, and 3 can be represented as $f(x) = (x+2)(x-1)(x-3)$. Expanding this, we get:
$$f(x) = x^3 - 2x^2 - 5x + 6.$$

49) To simplify $(2a^2b)^3(4ab^2)^2$, first expand each term:

$(2a^2b)^3 = 8a^6b^3,$
$(4ab^2)^2 = 16a^2b^4.$

Then multiply these results together:
$$8a^6b^3 \times 16a^2b^4 = 128a^{(6+2)}b^{(3+4)} = 128a^8b^7.$$

Therefore, the simplified expression is $128a^8b^7$.

50) The zero of the function g is the point where the graph crosses the x-axis. In this graph, g crosses the x-axis at $x = -2$. Therefore, the zero of g is -2.

7. Practice Test 6

CBEST Math Practice Test

Total number of questions: 50

Total time: 90 Minutes

Calculators are prohibited for the CBEST exam.

7.1 Practices

1) A cyclist covers distances of 30 km, 35 km, 40 km, 25 km, and 45 km during five consecutive hours. If the cyclist continues at an average speed of 55 km per hour for the next five hours, what is the total distance covered in 10 hours?

☐ A. 375 km

☐ B. 400 km

☐ C. 450 km

□ D. 500 km

□ E. 800 km

2) A truck needs to complete a journey of 360 miles. How long will the journey take if the truck travels at an average speed of 60 miles per hour (mph)?

   □ A. 5 hours

   □ B. 6 hours

   □ C. 7 hours

   □ D. 7 hours and 30 minutes

   □ E. 8 hours

3) What is the difference between the largest 4-digit number and the smallest 4-digit number?

   □ A. 8999

   □ B. 9000

   □ C. 9899

   □ D. 9999

   □ E. 10000

4) A spinner is divided into 8 equal sections labeled from A to H. It starts on section A and rotates clockwise, moving through one section every 2 minutes. To which section will the spinner point after 96 minutes?

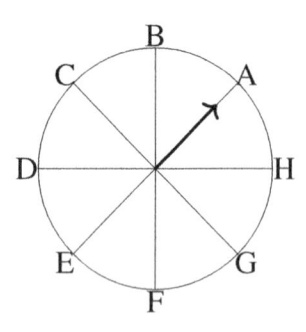

   □ A. A

   □ B. B

   □ C. C

   □ D. D

   □ E. E

7.1 Practices

5) In right triangle DEF, the lengths of the two legs are 8 *cm* (DE) and 15 *cm* (DF). What is the length of the hypotenuse (EF)?

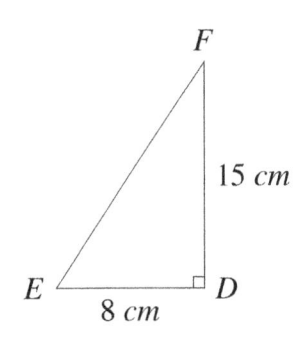

- ☐ A. 10 *cm*
- ☐ B. 17 *cm*
- ☐ C. 19 *cm*
- ☐ D. 23 *cm*
- ☐ E. 25 *cm*

6) The ratio of blue to green pencils in a box is 3 : 4. If there are 210 pencils in the box, how many blue pencils are there?

- ☐ A. 90
- ☐ B. 105
- ☐ C. 120
- ☐ D. 135
- ☐ E. 150

7) Solve for y: $5(y-2) = 3(y+3)+15$

- ☐ A. 10
- ☐ B. 7
- ☐ C. 3
- ☐ D. 17
- ☐ E. −7

8) What is 2.5% of 800?

- ☐ A. 20

- ☐ B. 40
- ☐ C. 60
- ☐ D. 80
- ☐ E. 100

9) Expand and simplify $(4x+3y)(6x+5y) =$
   - ☐ A. $24x^2 + 38xy + 15y^2$
   - ☐ B. $12x^2 + 27xy + 8y^2$
   - ☐ C. $10x^2 + 18xy + 6y^2$
   - ☐ D. $8x^2 + 30xy + 15y^2$
   - ☐ E. $6x^2 + 22xy + 15y^2$

10) Which of the following expressions is equivalent to $6x(3+4y)$?
    - ☐ A. $3x + 12xy$
    - ☐ B. $6x + 24xy$
    - ☐ C. $18x + 6xy$
    - ☐ D. $18x + 24xy$
    - ☐ E. $9x + 18xy$

11) If $z = 4cd + 2d^2$, what is z when $c = 3$ and $d = 4$?
    - ☐ A. 48
    - ☐ B. 56
    - ☐ C. 64
    - ☐ D. 80
    - ☐ E. 72

12) 12 is what percent of 48?
    - ☐ A. 10%
    - ☐ B. 15%
    - ☐ C. 25%
    - ☐ D. 30%
    - ☐ E. 50%

7.1 Practices

13) The perimeter of a trapezoid is 60 cm. If the lengths of the parallel sides are 15 cm and 25 cm, and the non-parallel sides are both 10 cm, what is its area?

☐ A. 100 cm^2

☐ B. 120 cm^2

☐ C. $25\sqrt{75}$ cm^2

☐ D. 160 cm^2

☐ E. $20\sqrt{75}$ cm^2

14) Three fourths of 20 is equal to $\frac{3}{5}$ of what number?

☐ A. 15

☐ B. 25

☐ C. 30

☐ D. 45

☐ E. 60

15) A refrigerator is originally priced at \$E. Its price is first increased by 10% in March and then decreased by 20% in April. What is the final price of the refrigerator in terms of \$E?

☐ A. $0.72\ E$

☐ B. $0.88\ E$

☐ C. $0.90\ E$

☐ D. $1.08\ E$

☐ E. $1.10\ E$

16) Determine the median of the following numbers: 2, 7, 12, 9, 14, 16, 3

☐ A. 7

☐ B. 9

☐ C. 12

☐ D. 14

☐ E. 17

17) A cylindrical water tank has a radius of 5 inches and a height of 10 inches. Calculate the total surface area of this water tank.

- [ ] A. 150π in^2
- [ ] B. 225π in^2
- [ ] C. 300π in^2
- [ ] D. 375π in^2
- [ ] E. 450π in^2

18) The average of 8, 12, 16, and y is 14. What is the value of y?

- [ ] A. 10
- [ ] B. 14
- [ ] C. 18
- [ ] D. 20
- [ ] E. 26

19) The price of a laptop was increased by 20% to $720. What was its original price?

- [ ] A. $600
- [ ] B. $650
- [ ] C. $680
- [ ] D. $700
- [ ] E. $750

20) What are the zeros of the function $g(x) = x^2 - 3x - 4$?

- [ ] A. 1
- [ ] B. $-1, 4$
- [ ] C. 1, 4
- [ ] D. $-4, 1$
- [ ] E. $-1, -4$

21) The area of a circle is 64π. What is the circumference of the circle?

- [ ] A. 8π
- [ ] B. 16π
- [ ] C. 32π
- [ ] D. 64π

7.1 Practices

☐ E. 128π

22) A $120 jacket, now selling for $72, is discounted by what percent?

☐ A. 20%

☐ B. 30%

☐ C. 40%

☐ D. 50%

☐ E. 60%

23) In 2005, the average worker's income increased $1500 per year, starting from a $30000 annual salary. Which equation represents income greater than the average? (I = income, x = number of years after 2005)

☐ A. $I > 1500x + 30000$

☐ B. $I > -1500x + 30000$

☐ C. $I < -1500x + 30000$

☐ D. $I < 1500x - 30000$

☐ E. $I < 28500x + 30000$

24) From last year, the price of gasoline has increased from $2.00 per gallon to $2.50 per gallon. The new price is what percent of the original price?

☐ A. 105%

☐ B. 115%

☐ C. 125%

☐ D. 135%

☐ E. 150%

25) A boat sails 40 miles north and then 30 miles west. How far is the boat from its starting point?

☐ A. 20 miles

☐ B. 30 miles

☐ C. 50 miles

☐ D. 70 miles

☐ E. 100 miles

26) Which of the following could be the product of two consecutive prime numbers?

☐ A. 15

☐ B. 21

☐ C. 25

☐ D. 33

☐ E. 55

27) Emma purchased a bicycle for $720. The bicycle is regularly priced at $800. What was the percent discount Emma received on the bicycle?

☐ A. 5%

☐ B. 10%

☐ C. 15%

☐ D. 20%

☐ E. 25%

28) The score of Olivia was one-third as much as that of Mia and the score of Lily was twice that of Mia. If the score of Lily was 60, what is the score of Olivia?

☐ A. 10

☐ B. 15

☐ C. 20

☐ D. 25

☐ E. 30

29) A bag contains 18 balls: two green, five black, eight blue, one brown, one red, and one white. If 17 balls are removed from the bag at random, what is the probability that a brown ball has been removed?

☐ A. $\frac{1}{10}$

☐ B. $\frac{1}{9}$

☐ C. $\frac{1}{6}$

☐ D. $\frac{16}{11}$

☐ E. $\frac{17}{18}$

30) The average of four consecutive even numbers is 13. What is the smallest number?

☐ A. 12

☐ B. 20

☐ C. 10

☐ D. 16

☐ E. 8

31) The price of a computer was $1200 in 2015. In 2017, the price of that computer was $900. What was the rate of depreciation of the price of the computer per year?

☐ A. 10%

☐ B. 12.5%

☐ C. 20%

☐ D. 25%

☐ E. 30%

32) The width of a box is one-third of its length. The height of the box is one-third of its width. If the length of the box is 36 cm, what is the volume of the box?

☐ A. 81 cm^3

☐ B. 162 cm^3

☐ C. 243 cm^3

☐ D. 1728 cm^3

☐ E. 1880 cm^3

33) If 40% of X is 20% of Y, then Y is what percent of X?

☐ A. 25%

☐ B. 50%

☐ C. 100%

☐ D. 200%

☐ E. 400%

34) How many possible pizza combinations can be made from four types of crust, three types of sauce, and five toppings?

☐ A. 12

☐ B. 24

☐ C. 30

☐ D. 60

☐ E. 72

35) A building 60 feet tall casts a shadow 30 feet long. Sarah is 5 feet tall. How long is Sarah's shadow?

☐ A. 7.5 feet

☐ B. 10 feet

☐ C. 2.5 feet

☐ D. 15 feet

☐ E. 20 feet

36) When a number is subtracted from 42 and the difference is divided by that number, the result is 5. What is the value of the number?

☐ A. 4

☐ B. 7

☐ C. 6

☐ D. 8

☐ E. 9

37) An angle is equal to one-eighth of its supplement. What is the measure of that angle?

☐ A. 8

☐ B. 20

☐ C. 36

☐ D. 45

☐ E. 72

38) Emma drove her car at a certain speed and covered a distance of 180 km in 5 hours. On the other hand, Olivia drove her car at a different speed and covered a distance of 210 km in 6 hours. What is the ratio of the average speed of Emma to the average speed of Olivia?

☐ A. 36 : 35

☐ B. 32 : 33

☐ C. 35 : 36

7.1 Practices

☐ D. 37 : 38

☐ E. 38 : 37

39) In a school, 6% of the students are members of the chess club, and 30% of the chess club members are also members of the math club. What percent of the students are members of both the chess club and the math club?

☐ A. 1.8%

☐ B. 2.0%

☐ C. 5.4%

☐ D. 18%

☐ E. 8%

40) Determine the value of x and y in the following system of linear equations:

$$\begin{cases} 2x + y = 6 \\ 3x - 2y = -5 \end{cases}$$

☐ A. $x = 3, y = 4$

☐ B. $x = 1, y = 4$

☐ C. $x = 3, y = -4$

☐ D. $x = -3, y = 4$

☐ E. $x = 0, y = 4$

41) A rectangular swimming pool measuring 10 m by 15 m needs to be covered with square tiles, each measuring 5 m². How many tiles are required to cover the entire pool floor?

☐ A. 30

☐ B. 60

☐ C. 90

☐ D. 120

☐ E. 150

42) A wire weighs 450 grams per meter. What is the weight in kilograms of 20 meters of this wire? (1 kilogram = 1000 grams)

☐ A. 0.9

☐ B. 9

☐ C. 90

☐ D. 900

☐ E. 9000

43) A fruit drink contains 5% fruit juice. If there are 15 ml of fruit juice, what is the total volume of the drink?

☐ A. 150 ml

☐ B. 300 ml

☐ C. 500 ml

☐ D. 1000 ml

☐ E. 3000 ml

44) In a sports team, the average height of 10 female players is 165 cm and the average height of 15 male players is 175 cm. What is the average height of all the 25 players in the team?

☐ A. 171.0 cm

☐ B. 171.5 cm

☐ C. 172.0 cm

☐ D. 173.0 cm

☐ E. 174.5 cm

45) The price of a smartphone is increased by 25% to $500. What was its price before the increase?

☐ A. $360

☐ B. $400

☐ C. $440

☐ D. $480

☐ E. $520

46) A credit union offers 3.2% simple interest on a fixed deposit. If you invest $5000, what will be the total interest earned in three years?

☐ A. $150

☐ B. $320

☐ C. $480

7.1 Practices

☐ D. $640

☐ E. $960

47) Multiply and express the product in scientific notation: $(3.2 \times 10^7) \times (4.5 \times 10^{-4})$.

☐ A. 144×10^3

☐ B. 1.44×10^3

☐ C. 14.4×10^4

☐ D. 1.44×10^4

☐ E. 14.4×10^3

48) If the height of a right cone is 10 cm and its base radius is 3 cm. What is its volume? (Use $\pi = 3$)

☐ A. 90 cm^3

☐ B. 94 cm^3

☐ C. 100 cm^3

☐ D. 120 cm^3

☐ E. 141 cm^3

49) If triangle *DEF* is reflected over the *y*-axis, what are the coordinates of the new image? In a coordinate plane, triangle *DEF* has coordinates: $(-1,4)$, $(-2,5)$, and $(5,9)$.

☐ A. $(-1,-4)$, $(-2,-5)$, $(-5,9)$

☐ B. $(1,4)$, $(2,5)$, $(5,-9)$

☐ C. $(-1,4)$, $(-2,5)$, $(5,9)$

☐ D. $(2,-1)$, $(5,-3)$, $(-3,2)$

☐ E. $(1,4)$, $(2,5)$, $(-5,9)$

50) In the right triangle shown below, calculate the value of *x*, the length of the hypotenuse.

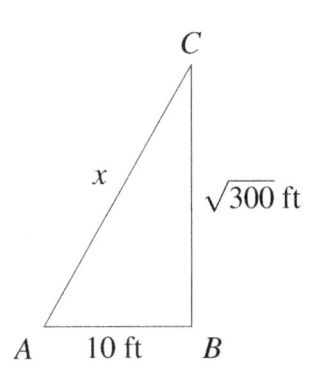

- [ ] A. 8 ft
- [ ] B. 10 ft
- [ ] C. 16 ft
- [ ] D. 18 ft
- [ ] E. 20 ft

7.2 Answer Keys

1) C. 450 km
2) B. 6 hours
3) A. 8999
4) A. A
5) B. 17 cm
6) A. 90
7) D. 17
8) A. 20
9) A. $24x^2 + 38xy + 15y^2$
10) D. $18x + 24xy$
11) D. 80
12) C. 25%
13) E. $20\sqrt{75}$ cm^2
14) B. 25
15) B. 0.88 E
16) B. 9
17) A. 150π in^2
18) D. 20
19) A. $600
20) B. $-1, 4$
21) B. 16π
22) C. 40%
23) A. $I > 1500x + 30000$
24) C. 125%
25) C. 50 miles

26) A. 15
27) B. 10%
28) A. 10
29) E. $\frac{17}{18}$
30) C. 10
31) B. 12.5%
32) D. 1728 cm^3
33) D. 200%
34) D. 60
35) C. 2.5 feet
36) B. 7
37) B. 20
38) A. 36 : 35
39) A. 1.8%
40) B. $x = 1, y = 4$
41) A. 30
42) B. 9
43) B. 300 ml
44) A. 171.0 cm
45) B. $400
46) C. $480
47) D. 1.44×10^4
48) A. 90 cm^3
49) B. $(1,4), (2,5), (-5,9)$
50) E. 20 ft

7.3 Answers with Explanation

1) In the first five hours, the cyclist covers a total of $30 + 35 + 40 + 25 + 45 = 175$ km. For the next five hours, at an average speed of 55 km per hour, the total distance is $5 \times 55 = 275$ km. The total distance covered in 10 hours is $175 + 275 = 450$ km. Therefore, the correct answer is C.

2) To find the time taken for the journey, divide the total distance by the speed. Thus,

$$\text{time taken} = \frac{360 \text{ miles}}{60 \text{ mph}} = 6 \text{ hours}.$$

Therefore, the correct answer is B.

3) The largest 4-digit number is 9999, and the smallest 4-digit number is 1000. The difference between these two numbers is calculated as $9999 - 1000 = 8999$. Therefore, the correct answer is A. 8999.

4) After 96 minutes, the spinner completes $\frac{96}{2} = 48$ moves. Since there are 8 sections, every 8 moves the spinner completes a full cycle and returns to section A. Therefore, after 48 moves, the spinner will have completed 6 full cycles (48 divided by 8) and will point to section A again.

5) Using the Pythagorean theorem for right triangle DEF, we calculate the length of the hypotenuse (EF) as $\sqrt{8^2 + 15^2} = \sqrt{64 + 225} = \sqrt{289} = 17$ cm. Therefore, the correct answer is B.

6) The total number of parts in the ratio is $3 + 4 = 7$. Each part represents $\frac{210}{7} = 30$ pencils. Therefore, the number of blue pencils, which is 3 parts, is $3 \times 30 = 90$. Thus, the correct answer is A.

7) Start by expanding both sides: $5y - 10 = 3y + 9 + 15$. Simplifying further, we get $5y - 10 = 3y + 24$. Subtract $3y$ from both sides to get $2y - 10 = 24$. Adding 10 to both sides gives $2y = 34$. Dividing by 2 yields $y = 17$. Therefore, the correct answer is D.

8) To find 2.5% of 800, convert the percentage to a decimal and multiply by the base value: $0.025 \times 800 = 20$. Therefore, the correct answer is A.

9) To expand $(4x + 3y)(6x + 5y)$, apply the distributive property (FOIL method): $4x \times 6x = 24x^2$, $4x \times 5y = 20xy$, $3y \times 6x = 18xy$, $3y \times 5y = 15y^2$. Combining like terms gives $24x^2 + 38xy + 15y^2$. Therefore, the correct

7.3 Answers with Explanation

answer is A.

10) To simplify $6x(3+4y)$, distribute $6x$ across the terms inside the parenthesis: $(6x)(3) = 18x$, and $(6x)(4y) = 24xy$. Therefore, the expression simplifies to $18x + 24xy$, which matches option D.

11) Substitute $c = 3$ and $d = 4$ into the equation $z = 4cd + 2d^2$ to get

$$z = (4)(3)(4) + 2(4^2) = 48 + 32 = 80.$$

Therefore, the correct answer is D.

12) To determine what percent 12 is of 48, we use the formula for calculating percentage:

$$\text{Percentage} = \left(\frac{\text{Part}}{\text{Whole}}\right) \times 100\%.$$

In this problem, 12 is the "Part" and 48 is the "Whole". Thus, substituting these values into the formula, we get:

$$\text{Percentage} = \left(\frac{12}{48}\right) \times 100\%.$$

Simplifying the fraction $\frac{12}{48}$ gives $\frac{1}{4}$, as 12 is one-fourth of 48. Converting the fraction to a decimal, $\frac{1}{4} = 0.25$. Now, multiplying this decimal by 100% gives the percentage:

$$0.25 \times 100\% = 25\%.$$

13)

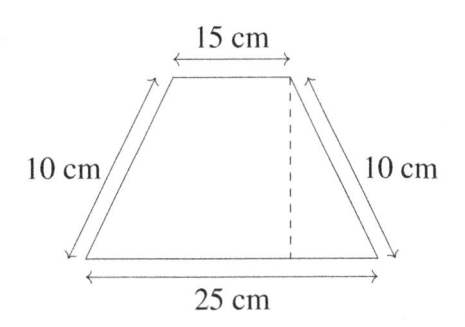

To find the area of the trapezoid, first determine the height using the Pythagorean theorem. The difference between the lengths of the parallel sides is 10 cm, half of which is 5 cm. The height h is then calculated as

$\sqrt{10^2 - 5^2} = \sqrt{75}$ cm. The area of the trapezoid is then

$$\frac{1}{2}(15+25)\sqrt{75} = 20\sqrt{75} \text{ cm}^2.$$

14) Let the unknown number be x. Then, $(\frac{3}{4})(20) = (\frac{3}{5})(x) \Rightarrow 15 = (\frac{3}{5})(x)$. Solving for x, we get $x = 15 \div \frac{3}{5} = 15 \times \frac{5}{3} = 25$. Therefore, the correct answer is B.

15) First, calculate the price after a 10% increase:

$$E + 10\% \text{ of } E = E + 0.10E = 1.10E.$$

Then, calculate the price after a 20% decrease:

$$1.10E - 20\% \text{ of } 1.10E = 1.10E - 0.20 \times 1.10E = 1.10E - 0.22E = 0.88E.$$

Therefore, the final price is $0.88E$.

16) First, arrange the numbers in ascending order: 2, 3, 7, 9, 12, 14, 16. With seven numbers, the median is the fourth number in the sequence, which is 9. Therefore, the median of these numbers is 9.

17) The formula for the surface area of a cylinder is $2\pi r^2 + 2\pi rh = 2\pi r(r+h)$. Here, the radius r is 5 inches, and the height h is 10 inches. Therefore, the surface area is $2 \times \pi \times 5 \times (5+10) = 150\pi$ in². Thus, option A is correct.

18) The average of four numbers is the sum of the numbers divided by 4. Therefore, $\frac{8+12+16+y}{4} = 14$. Simplifying, $36 + y = 56$. Thus, $y = 56 - 36 = 20$. The value of y is 20.

19) Let the original price of the laptop be x dollars. After a 20% increase, the new price is $720. This can be expressed as $x + (20\% \text{ of } x) = \$720 \Rightarrow x + 0.20x = \$720$. Simplifying, $1.20x = \$720$. Therefore, the original price x is $\frac{\$720}{1.20} = \$600$.

20) To find the zeros of the function $g(x) = x^2 - 3x - 4$, set the function equal to zero: $x^2 - 3x - 4 = 0$. Solve this quadratic equation by factoring: $(x-4)(x+1) = 0$. The zeros are the solutions to $x - 4 = 0$ and $x + 1 = 0$, which are $x = 4$ and $x = -1$ respectively. Therefore, the zeros of the function are -1 and 4.

7.3 Answers with Explanation

21) First, find the radius r of the circle. The area A is given by $A = \pi r^2$, so $64\pi = \pi r^2$. Solving for r, we find $r = 8$. The circumference C of a circle is given by $C = 2\pi r$, so $C = (2\pi)(8) = 16\pi$. Therefore, the correct answer is B.

22) The discount is calculated by the formula:

$$\text{Discount} = \frac{\text{Original Price} - \text{Sale Price}}{\text{Original Price}} \times 100\%.$$

So,

$$\text{Discount} = \frac{120 - 72}{120} \times 100\% = \frac{48}{120} \times 100\% = 40\%.$$

Therefore, the correct answer is C.

23) The income I increases by $1500 each year after 2005, starting from $30000. Thus, the income in any year after 2005 can be represented by the equation $I = 1500x + 30000$, where x is the number of years after 2005. To represent income greater than the average, the inequality $I > 1500x + 30000$ is used. Therefore, the correct answer is A.

24) To find what percent the new price is of the original price use the percent formula:

$$\text{part} = \frac{\text{percent}}{100} \times \text{whole}.$$

So we have:

$$2.50 = \frac{\text{percent}}{100} \times 2.00 \Rightarrow 2.50 = \frac{\text{percent} \times 2.00}{100} \Rightarrow 250 = \text{percent} \times 2.00 \Rightarrow \text{percent} = \frac{250}{2.00} = 125\%.$$

Therefore, the correct answer is C.

25) The distance from the starting point can be found using the Pythagorean theorem. The boat's journey forms a right triangle with legs of 40 miles and 30 miles. The distance d from the starting point is the hypotenuse of this triangle, so

$$d = \sqrt{40^2 + 30^2} = \sqrt{1600 + 900} = \sqrt{2500} = 50 \text{ miles}.$$

Therefore, the correct answer is C.

26) Of the given options, 15 is the only number that is the product of two consecutive prime numbers, which are 3 and 5 (as $15 = 3 \times 5$). Therefore, the correct answer is A.

27) To calculate the percent discount, we use the formula:

$$\text{Percent Discount} = \frac{\text{Regular Price} - \text{Sale Price}}{\text{Regular Price}} \times 100\%.$$

In this case, it is $\frac{800-720}{800} \times 100\% = \frac{80}{800} \times 100\% = 10\%$. Therefore, the correct answer is B.

28) If the score of Lily is twice that of Mia and is 60, then the score of Mia is $\frac{60}{2} = 30$. If the score of Olivia is one-third as much as that of Mia, then the score of Olivia is $\frac{30}{3} = 10$. Therefore, the correct answer is A.

29) The probability of not drawing a brown ball can be calculated by first finding the number of ways to choose 17 balls from the 17 non-brown balls (since there's only one brown ball), which is $\binom{17}{17}$. The total number of ways to choose 17 balls from 18 is $\binom{18}{17}$. Therefore, the probability of not drawing a brown ball is $\frac{\binom{17}{17}}{\binom{18}{17}} = \frac{1}{18}$. The probability of drawing at least one brown ball is then $1 - \frac{1}{18}$, which is $\frac{17}{18}$. Hence, the correct answer is E.

30) Let the four consecutive even numbers be x, $x+2$, $x+4$, and $x+6$. Their average is given by

$$\frac{x + (x+2) + (x+4) + (x+6)}{4} = 13.$$

Simplifying, we have $\frac{4x+12}{4} = 13$, leading to $4x + 12 = 52$. Solving for x gives $4x = 40$, and $x = 10$.

31) The rate of depreciation can be calculated using the formula:

$$\text{Rate of Depreciation} = \frac{\text{Initial Price} - \text{Final Price}}{\text{Initial Price}} \times \frac{1}{\text{Number of Years}} \times 100\%.$$

In this case, it is

$$(\frac{1200 - 900}{1200})(\frac{1}{2}) \times 100\% = (\frac{300}{1200})(\frac{1}{2}) \times 100\% = 12.5\%.$$

Therefore, the correct answer is B. 12.5%.

7.3 Answers with Explanation

32) Given that the width of the box is one-third of its length, we can calculate the width as follows:

$$\text{Width} = \frac{1}{3} \times \text{Length} = \frac{1}{3} \times 36 \text{ cm} = 12 \text{ cm}.$$

Next, the height of the box is one-third of its width:

$$\text{Height} = \frac{1}{3} \times \text{Width} = \frac{1}{3} \times 12 \text{ cm} = 4 \text{ cm}.$$

Now, we can calculate the volume of the box using the formula for the volume of a rectangular box:

$$\text{Volume} = \text{Length} \times \text{Width} \times \text{Height} = 36 \text{ cm} \times 12 \text{ cm} \times 4 \text{ cm} = 1728 \text{ cm}^3.$$

Therefore, the volume of the box is 1728 cm^3, which corresponds to option D.

33) Given that 40% of X is 20% of Y, we can write this as an equation:

$$0.4X = 0.2Y \Rightarrow Y = \frac{0.4}{0.2}X \Rightarrow Y = 2X.$$

Therefore, Y is 200% of X, which corresponds to option D.

34) To find the number of possible pizza combinations, we can multiply the number of choices for each component:

Number of crust choices: 4

Number of sauce choices: 3

Number of topping choices: 5

Total combinations = $4 \times 3 \times 5 = 60$

So, there are 60 possible pizza combinations, which corresponds to option D.

35) We can use the concept of similar triangles to find the length of Sarah's shadow. The ratio of the height of the building to its shadow length is the same as the ratio of Sarah's height to her shadow length:

$$\frac{\text{Height of building}}{\text{Shadow length of building}} = \frac{\text{Height of Sarah}}{\text{Shadow length of Sarah}}.$$

Substituting the given values:

$$\frac{60 \text{ feet}}{30 \text{ feet}} = \frac{5 \text{ feet}}{\text{Shadow length of Sarah}}.$$

Solving for the shadow length of Sarah:

$$\text{Shadow length of Sarah} = \frac{5 \text{ feet} \times 30 \text{ feet}}{60 \text{ feet}} = 2.5 \text{ feet}.$$

So, Sarah's shadow is 2.5 feet, which corresponds to option C.

36) Let the number be x. According to the given information:

$$\frac{42 - x}{x} = 5.$$

Now, solve for x:

$$42 - x = 5x \Rightarrow 42 = 6x \Rightarrow x = \frac{42}{6} = 7.$$

So, the value of the number is 7, which corresponds to option B.

37) Let the measure of the angle be x degrees. According to the given information, the angle is equal to one-eighth of its supplement, which can be represented as:

$$x = \left(\frac{1}{8}\right)(180 - x)$$

Now, solve for x:

$$8x = 180 - x \Rightarrow 9x = 180 \Rightarrow x = \frac{180}{9} = 20.$$

So, the measure of the angle is 20 degrees, which corresponds to option B.

38) To find the average speed, divide the total distance by the total time taken.

For Emma: Average speed = $\frac{180 \text{ km}}{5 \text{ hours}} = 36$ km/h

For Olivia: Average speed = $\frac{210 \text{ km}}{6 \text{ hours}} = 35$ km/h

The ratio of Emma's average speed to Olivia's average speed is $36 : 35$.

39) To find the percent of students who are members of both the chess club and the math club, we can multiply the percentages. Percent of students in the chess club and math club = $6\% \times 30\% = 0.06 \times 0.30 = 0.018 = 1.8\%$.

7.3 Answers with Explanation 173

So, 1.8% of the students are members of both clubs.

40) To solve the system of equations $\begin{cases} 2x+y=6, \\ 3x-2y=-5 \end{cases}$ we first express y from the first equation: $y = 6-2x$. Substituting this into the second equation gives: $3x-2(6-2x) = -5$. Expanding and rearranging terms, we get: $3x-12+4x = -5$, which simplifies to $7x = 7$. Solving for x, we find $x = 1$. Now substituting $x = 1$ into $y = 6-2x$ gives: $y = 6-2 \times 1 = 4$. Therefore, $y = 4$.

41) First, calculate the area of the pool: $10 \text{ m} \times 15 \text{ m} = 150 \text{ m}^2$. Each tile covers 5 m^2, so the number of tiles needed is $\frac{150 \text{ m}^2}{5 \text{ m}^2} = 30$ tiles.

42) First, calculate the total weight in grams: $450 \text{ grams/meter} \times 20 \text{ meters} = 9000$ grams. To convert this to kilograms, divide by 1000: $\frac{9000 \text{ grams}}{1000} = 9$ kilograms. Therefore, the weight of the wire is 9 kilograms.

43) To find the total volume of the drink, use the proportion of fruit juice to the entire drink. Since 5% of the drink is fruit juice, and there are 15 ml of fruit juice, the equation is

$$\frac{15 \text{ ml}}{\text{Total Volume}} = \frac{5}{100}.$$

Cross-multiplying and solving for the Total Volume gives Total Volume $= \frac{15 \times 100}{5} = 300$ ml. Therefore, the total volume of the drink is 300 ml.

44) First, find the total height of the female players: $10 \times 165 \text{ cm} = 1650$ cm. Next, find the total height of the male players:

$$15 \times 175 \text{ cm} = 2625 \text{ cm}.$$

Add these to get the total height of all players: $1650 \text{ cm} + 2625 \text{ cm} = 4275$ cm. Finally, divide by the total number of players to get the average: $\frac{4275 \text{ cm}}{25} = 171$ cm. Therefore, the average height of all players is 171.0 cm.

45) If the final price after a 25% increase is $500, then $500 is 125% of the original price. To find the original price, set up the equation:

$$\frac{125}{100} \times \text{Original Price} = \$500.$$

Solving for the original price gives

$$\text{Original Price} = \frac{\$500 \times 100}{125} = \$400.$$

Thus, the original price of the smartphone was $400.

46) Simple interest is calculated as

$$\text{Principal} \times \text{Interest Rate} \times \text{Time}.$$

Here, the principal is $5000, the rate is 3.2% or 0.032, and the time is 3 years. Therefore, the interest earned is

$$\$5,000 \times 0.032 \times 3 = \$480.$$

Thus, the total interest earned in three years is $480.

47) When multiplying numbers in scientific notation, multiply the coefficients and add the exponents. Here, multiply 3.2 by 4.5 to get 14.4, and add the exponents 7 and -4 to get 3. Thus, the product is 14.4×10^3. To express this in standard scientific notation, divide the coefficient by 10 and increase the exponent by 1, resulting in 1.44×10^4.

48) The formula for the volume of a cone is $\frac{1}{3}\pi r^2 h$, where r is the radius and h is the height. Substitute $r = 3$ cm and $h = 10$ cm into the formula to get

$$\frac{1}{3}\pi (3 \text{ cm})^2 (10 \text{ cm}) = 30\pi \text{ cm}^3 = 90 \text{ cm}^3.$$

49) When triangle DEF is reflected over the y-axis, the x-coordinates of its points change sign while the y-coordinates remain the same. Therefore, the new image has the following coordinates:

- Point D: $(-1,4)$ becomes $(1,4)$
- Point E: $(-2,5)$ becomes $(2,5)$
- Point F: $(5,9)$ becomes $(-5,9)$

So, the coordinates of the new image are $(1,4)$, $(2,5)$, $(-5,9)$, which corresponds to option B.

7.3 Answers with Explanation

50) In a right triangle, the length of the hypotenuse x can be found using the Pythagorean theorem:

$$x^2 = (10 \text{ ft})^2 + (\sqrt{300} \text{ ft})^2.$$

Simplifying, we get:

$$x^2 = 100 \text{ ft}^2 + 300 \text{ ft}^2 = 400 \text{ ft}^2.$$

Therefore, solving for x gives:

$$x = \sqrt{400 \text{ ft}^2} = 20 \text{ ft}.$$

Thus, the length of the hypotenuse x is 20 ft.

8. Practice Test 7

CBEST Math Practice Test

Total number of questions: 50

Total time: 90 Minutes

Calculators are prohibited for the CBEST exam.

8.1 Practices

1) The mean of 40 test scores was calculated as 85. However, it was later discovered that one of the scores was misread as 88 instead of 68. What is the correct mean?

☐ A. 82
☐ B. 83.5
☐ C. 84
☐ D. 84.5

8.1 Practices

☐ E. 85

2) Two dice are rolled simultaneously. What is the probability of getting a sum of 7 or 9?

☐ A. $\frac{1}{4}$

☐ B. $\frac{5}{36}$

☐ C. $\frac{1}{9}$

☐ D. $\frac{1}{12}$

☐ E. $\frac{5}{18}$

3) In the triangle shown below, two exterior angles are 105° and 125°. What is the value of the unknown angle y?

☐ A. 40°

☐ B. 50°

☐ C. 55°

☐ D. 60°

☐ E. 65°

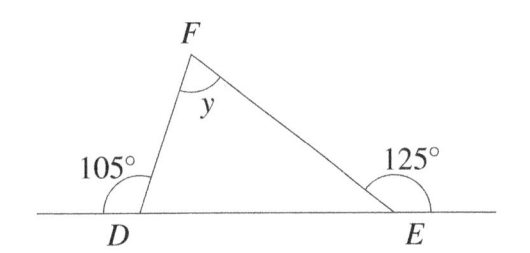

4) Which of the following expressions is equivalent to the one given below?

$$(3x - 4y)(x + 2y)$$

☐ A. $3x^2 - 8y^2$

☐ B. $3x^2 + 2xy - 8y^2$

☐ C. $3x^2 + 6xy - 8y^2$

☐ D. $3x^2 - 2xy - 8y^2$

☐ E. $5x^2 + 6xy - 4y^2$

5) Determine the product of all possible values of x that satisfy the given equation.

$$|2x - 8| = 6$$

☐ A. 1

☐ B. 2

- [ ] C. 5
- [ ] D. 7
- [ ] E. 11

6) What is the slope of a line that is perpendicular to the line defined by the equation:

$$6x + 2y = 10$$

- [ ] A. -3
- [ ] B. $-\frac{1}{3}$
- [ ] C. $\frac{1}{3}$
- [ ] D. 3
- [ ] E. $\frac{5}{3}$

7) Determine the value of the expression $4(x - 3y) + (3 - 2x)^2$ for $x = 2$ and $y = -1$.
- [ ] A. 5
- [ ] B. 13
- [ ] C. 17
- [ ] D. 19
- [ ] E. 21

8) A water tank has a volume of 3000 cubic meters. The tank is 20 meters in length and 15 meters in width. What is the depth of the water tank?
- [ ] A. 5 meters
- [ ] B. 8 meters
- [ ] C. 10 meters
- [ ] D. 12 meters
- [ ] E. 15 meters

9) For their party, the Smiths can choose between 3 appetizers, 4 main courses, and 3 desserts. How many different combinations of a three-course meal can they create?
- [ ] A. 10
- [ ] B. 24

8.1 Practices

- [ ] C. 30
- [ ] D. 36
- [ ] E. 40

10) If you have five one-foot rulers, how many students can you give $\frac{1}{4}$ of a ruler to?

- [ ] A. 5
- [ ] B. 10
- [ ] C. 20
- [ ] D. 25
- [ ] E. 40

11) What is the area of a rectangle whose length is 3 times its width and whose perimeter is 32?

- [ ] A. 48
- [ ] B. 72
- [ ] C. 96
- [ ] D. 108
- [ ] E. 144

12) In the right triangle shown below, sides *DE* and *EF* have lengths of 8 and 6 respectively. If sides *DE* and *EF* are both tripled in length, what will be the ratio of the new perimeter of the triangle to its new area?

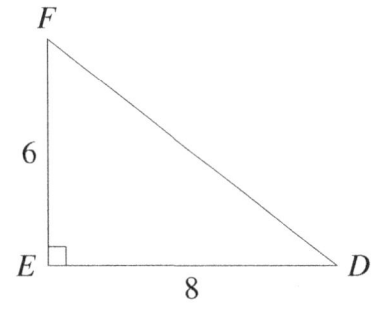

- [ ] A. $\frac{1}{3}$
- [ ] B. $\frac{5}{6}$
- [ ] C. $\frac{3}{4}$
- [ ] D. $\frac{3}{2}$
- [ ] E. 2

13) The average of seven numbers is 24. If an eighth number, 36, is added, what is the new average?

- [ ] A. 23
- [ ] B. 24.25
- [ ] C. 24.5
- [ ] D. 25.5
- [ ] E. 30

14) The ratio of cats to dogs in a pet store is 5 : 8. If there are 65 animals in total, how many more cats should be added to make the ratio 1 : 1?

☐ A. 5

☐ B. 13

☐ C. 15

☐ D. 20

☐ E. 40

15) Mrs. Smith saves $3,000 out of her monthly family income of $75,000. What fractional part of her income does she save?

☐ A. $\frac{1}{15}$

☐ B. $\frac{1}{20}$

☐ C. $\frac{1}{25}$

☐ D. $\frac{4}{75}$

☐ E. $\frac{4}{100}$

16) A school has allocated $30,000 for its annual sports festival. They have already spent $18,000 on uniforms. If each piece of sports equipment costs $120, which of the following inequalities represents the maximum number of equipment pieces they can buy?

☐ A. $120x + 18,000 \leq 30,000$

☐ B. $120x + 18,000 \geq 30,000$

☐ C. $18,000x + 120 \leq 30,000$

☐ D. $18,000x + 120 \geq 30,000$

☐ E. $18,000x + 18000 \geq 30,000$

17) To pass her chemistry class, Linda needs an average score of at least 75%. Her scores on the first four tests are 70%, 78%, 82%, and 75%. What is the minimum score she must achieve on her fifth test to pass the class?

☐ A. 75%

☐ B. 70%

☐ C. 65%

☐ D. 60%

☐ E. 55%

8.1 Practices

18) A plant grows $2\frac{1}{4}$ inches in $\frac{1}{3}$ of a year. What is its growth rate in inches per year?

☐ A. $6\frac{3}{4}$

☐ B. $6\frac{2}{3}$

☐ C. $5\frac{1}{2}$

☐ D. $4\frac{3}{4}$

☐ E. $3\frac{1}{3}$

19) A team is painting a fence. They can paint 20 *cm* of the fence per minute. After 60 minutes, $\frac{2}{3}$ of the fence is painted. What is the total length of the fence in meters?

☐ A. 5 *m*

☐ B. 15 *m*

☐ C. 18 *m*

☐ D. 27 *m*

☐ E. 30 *m*

20) Simplify the expression $5x^3y^2(3x^3y^2)^2$.

☐ A. $45x^6y^4$

☐ B. $45x^9y^6$

☐ C. $135x^6y^4$

☐ D. $135x^9y^6$

☐ E. $225x^9y^6$

21) Maria earns $40 an hour. Sarah earns 15% less than Maria. How much money does Sarah earn in an hour?

☐ A. $32.00

☐ B. $34.00

☐ C. $36.00

☐ D. $38.00

☐ E. $40.00

22) Last month, 18,000 people visited a museum. This month, four times as many people booked tickets, but one-fifth of them changed their plans. How many people are visiting the museum this month?

☐ A. 57,600

☐ B. 64,800

☐ C. 72,000

☐ D. 86,400

☐ E. 96,000

23) What is the perimeter of a square that has an area of 64 square inches?

☐ A. 128 inches

☐ B. 96 inches

☐ C. 64 inches

☐ D. 32 inches

☐ E. 16 inches

24) In rectangle $WXYZ$ with an area of 150 square units, point E is the midpoint of XY. A triangle with E as the top vertex is formed inside the rectangle, as shown in the diagram. What is the area of the shaded part of the rectangle?

☐ A. 50

☐ B. 75

☐ C. 80

☐ D. 85

☐ E. 95

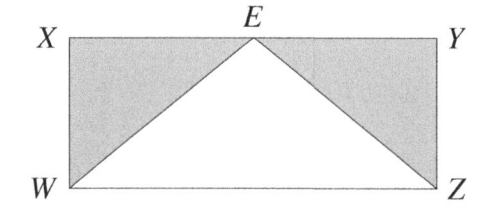

25) Set C contains all even integers from 20 to 200, inclusive, and Set D contains all multiples of 5 from 100 to 250, inclusive. How many integers are included in C but not in D?

☐ A. 80

☐ B. 75

☐ C. 90

☐ D. 65

☐ E. 91

26) Simplify the expression $(2x^3 - 3x^2 + 4x) + (7x^3 + 4x^2 - 6x)$.

☐ A. $5x^3 + x^2 - 2x$

☐ B. $7x^3 + x^2 - 10x$

☐ C. $9x^3 + x^2 - 2x$

8.1 Practices

- ☐ D. $9x^3 + x^2 + 10x$
- ☐ E. $9x^3 - x^2 - 2x$

27) If the ordered pair $(3, -7)$ is reflected over the y-axis, what is the new ordered pair?

- ☐ A. $(-3, -7)$
- ☐ B. $(3, 7)$
- ☐ C. $(-7, 3)$
- ☐ D. $(7, -3)$
- ☐ E. $(3, -7)$

28) The cube of a number is $\frac{8}{27}$. What is the square of that number?

- ☐ A. $\frac{4}{9}$
- ☐ B. $\frac{2}{3}$
- ☐ C. $\frac{16}{81}$
- ☐ D. $\frac{64}{729}$
- ☐ E. $\frac{32}{243}$

29) A freelance graphic designer charges $15 per hour of work. If he works 8 hours a day and spend $3 on software subscription fees for each hour worked, how much money does he make in one day?

- ☐ A. $96
- ☐ B. $104
- ☐ C. $120
- ☐ D. $132
- ☐ E. $94

30) A hiker starts at Base Camp B and travels 40 miles due south and then 30 miles due east. At this point, what is the shortest distance from the hiker to Base Camp B?

- ☐ A. 85 miles
- ☐ B. 65 miles
- ☐ C. 60 miles
- ☐ D. 55 miles
- ☐ E. 50 miles

31) What is the equivalent temperature of 95°F in Celsius using the conversion formula $C = \frac{5}{9}(F - 32)$?

☐ A. 30

☐ B. 35

☐ C. 37

☐ D. 38

☐ E. 40

32) In a basket, there are 20 red apples, 15 green apples, 25 bananas, and 40 oranges. What is the probability of randomly selecting a banana from the basket?

☐ A. $\frac{1}{4}$

☐ B. $\frac{1}{5}$

☐ C. $\frac{1}{10}$

☐ D. $\frac{1}{25}$

☐ E. $\frac{1}{100}$

33) Fill in the missing term in the given sequence:

$$4, 5, 7, 10, 14, 19, 25, \_\_\_, 40$$

☐ A. 26

☐ B. 28

☐ C. 30

☐ D. 31

☐ E. 32

34) The perimeter of a rectangular park is 96 meters. If the width is one-third of its length, what is the length of the park?

☐ A. 14 meters

☐ B. 24 meters

☐ C. 28 meters

☐ D. 36 meters

☐ E. 48 meters

8.1 Practices

35) How many positive integers satisfy the inequality $2x + 3 < 19$?

- ☐ A. 5
- ☐ B. 6
- ☐ C. 7
- ☐ D. 8
- ☐ E. 9

36) The table below shows the number of different colored books on a shelf:

Color	Number of Books
Red	15
Blue	25
Green	35

There are also yellow books on the shelf. Which of the following can NOT be the probability of randomly selecting a yellow book from the shelf?

- ☐ A. $\frac{1}{2}$
- ☐ B. $\frac{1}{4}$
- ☐ C. $\frac{5}{6}$
- ☐ D. $\frac{1}{5}$
- ☐ E. $\frac{2}{3}$

37) Determine the slope of the line represented by the equation $3x + 2y = 6$.

- ☐ A. -1.5
- ☐ B. -1
- ☐ C. 0.5
- ☐ D. 1
- ☐ E. 1.5

38) What is the volume of a container with the following dimensions?

Height = 4 cm, Width = 7 cm, Length = 8 cm

- ☐ A. $28 \; cm^3$
- ☐ B. $112 \; cm^3$

- C. 224 cm^3
- D. 256 cm^3
- E. 560 cm^3

39) Simplify the expression. $(6x^3+9x^2-4x^4)-(5x^2-3x^4+7x^3)$
- A. $x^4+6x^3+14x^2$
- B. $-x^4-x^3+4x^2$
- C. $-7x^4+13x^3+4x^2$
- D. $7x^4+13x^3-4x^2$
- E. $x^4-x^3+14x^2$

40) In two successive years, the sales of a company increased by 15% and then by 25%. What is the total percentage increase in the sales after two years?
- A. 40%
- B. 42.5%
- C. 43.75%
- D. 50.25%
- E. 60%

41) Simplify the expression. $(x^4)(x^5)$
- A. x^1
- B. x^9
- C. x^{16}
- D. x^{20}
- E. x^{25}

42) Given that the total sum of six distinct negative integers is -53 and the smallest integer among them is -12, determine the maximum possible value for the remaining five integers.
- A. -12
- B. -11
- C. -8
- D. -4

8.1 Practices

- ☐ E. −3

43) A certain quantity of a substance is increased by 15% to improve its efficacy. If the new amount is 115 grams, what was the original quantity?

- ☐ A. 92 grams
- ☐ B. 95 grams
- ☐ C. 100 grams
- ☐ D. 105 grams
- ☐ E. 110 grams

44) A flagpole stands next to a building, casting a shadow of 9 meters when a 1.5-meter stick next to it casts a shadow of 1 meter. How tall is the flagpole?

- ☐ A. 6 meters
- ☐ B. 9 meters
- ☐ C. 10.5 meters
- ☐ D. 12 meters
- ☐ E. 13.5 meters

45) Paul left a $18.00 tip for a meal that cost $90.00. What was the percentage of the tip relative to the cost of the meal?

- ☐ A. 15%
- ☐ B. 18%
- ☐ C. 20%
- ☐ D. 22%
- ☐ E. 25%

46) If 40% of a number is 16, what is the number?

- ☐ A. 10
- ☐ B. 20
- ☐ C. 30
- ☐ D. 40
- ☐ E. 50

47) If C is one-third of D and D is 24, what is the value of C?

☐ A. 8

☐ B. 6

☐ C. 4

☐ D. 12

☐ E. 16

48) If Lily is 15 kilometers ahead of Sam, who is cycling at 12 kilometers per hour, and Sam cycles at a speed of 15 kilometers per hour, how long will it take Sam to catch up with Lily?

☐ A. 2 hours

☐ B. 3 hours

☐ C. 4 hours

☐ D. 5 hours

☐ E. 6 hours

49) In a class of 60 students, 18 failed the final exam. What percentage of the students passed the exam?

☐ A. 65%

☐ B. 70%

☐ C. 75%

☐ D. 80%

☐ E. 85%

50) Given a line segment *PQ* of length 60 *cm*, with point *R* dividing it into two segments such that *PR* is twice as long as *RQ*, determine the length of segment *RQ*.

☐ A. 40 *cm*

☐ B. 20 *cm*

☐ C. 15 *cm*

☐ D. 10 *cm*

☐ E. 5 *cm*

8.2 Answer Keys

1) D. 84.5
2) E. $\frac{5}{18}$
3) B. 50°
4) B. $3x^2 + 2xy - 8y^2$
5) D. 7
6) C. $\frac{1}{3}$
7) E. 21
8) C. 10 meters
9) D. 36
10) C. 20
11) A. 48
12) A. $\frac{1}{3}$
13) D. 25.5
14) C. 15
15) C. $\frac{1}{25}$
16) A. $120x + 18,000 \leq 30,000$
17) B. 70%
18) A. $6\frac{3}{4}$
19) C. 18 m
20) B. $45x^9y^6$
21) B. $34.00
22) A. 57,600
23) D. 32 inches
24) B. 75
25) A. 80
26) C. $9x^3 + x^2 - 2x$
27) A. $(-3, -7)$
28) A. $\frac{4}{9}$
29) A. $96
30) E. 50 miles
31) B. 35
32) A. $\frac{1}{4}$
33) E. 32
34) D. 36 meters
35) C. 7
36) D. $\frac{1}{5}$
37) A. -1.5
38) C. 224 cm^3
39) B. $-x^4 - x^3 + 4x^2$
40) C. 43.75%
41) B. x^9
42) E. -3
43) C. 100 grams
44) E. 13.5 meters
45) C. 20%
46) D. 40
47) A. 8
48) D. 5 hours
49) B. 70%
50) B. 20 cm

8.3 Answers with Explanation

1) The incorrect total score was $40 \times 85 = 3400$. The difference in the misread score is $88 - 68 = 20$. Subtracting this from the total gives the correct total score $3400 - 20 = 3380$. The correct mean is then:

$$\frac{3380}{40} = 84.5.$$

Therefore, the correct mean is 84.5, making option D correct.

2) The possible ways to get a sum of 7 are $(1,6), (2,5), (3,4), (4,3), (5,2), (6,1)$, totaling 6 combinations. For a sum of 9, the combinations are $(3,6), (4,5), (5,4), (6,3)$, totaling 4 combinations. There are 36 possible outcomes when rolling two dice. The probability is:

$$\frac{6+4}{36} = \frac{10}{36} = \frac{5}{18}.$$

Therefore, the correct answer is E.

3) The exterior angle of a triangle is equal to the sum of non-adjacent interior angles. Consider the angle $125°$ as an exterior angle, we have:

$$y + (180 - 105) = 125 \Rightarrow y = 125 - 75 = 50.$$

Therefore, the value of y is $50°$, making option B correct.

4) Expanding the binomials, we have:

$$(3x - 4y)(x + 2y) = 3x^2 + 6xy - 4xy - 8y^2 = 3x^2 + 2xy - 8y^2.$$

Therefore, the correct expansion is $3x^2 + 2xy - 8y^2$, making option B correct.

5) We solve the absolute value equation:

$$2x - 8 = 6 \quad \text{or} \quad 2x - 8 = -6.$$

8.3 Answers with Explanation

This gives us $x = 7$ or $x = 1$. The product of these values is:

$$7 \times 1 = 7.$$

Therefore, the correct product is 7, making option D correct.

6) First, rewrite the equation in slope-intercept form to find the slope of the original line:

$$6x + 2y = 10 \Rightarrow 2y = -6x + 10 \Rightarrow y = -3x + 5.$$

The slope of this line is -3. The slope of a line perpendicular to this line is the negative reciprocal of -3, which is $\frac{1}{3}$. Therefore, the correct answer is C.

7) Substituting the values of x and y, we have:

$$4(2 - 3(-1)) + (3 - 2(2))^2 = 4(2 + 3) + (3 - 4)^2 = 4(5) + (-1)^2 = 20 + 1 = 21.$$

Therefore, the correct value is 21, making option E correct.

8) The volume of the tank is given by the product of its length, width, and depth. To find the depth, we divide the volume by the product of the length and width:

$$\text{Depth} = \frac{3000}{20 \times 15} = 10 \text{ meters}.$$

Therefore, the correct depth is 10 meters, making option C correct.

9) Using the counting principle, the total number of different combinations is the product of the number of choices for each course:

$$3 \times 4 \times 3 = 36.$$

Therefore, there are 36 different combinations, making option D correct.

10) Each one-foot ruler can be divided into 4 parts of $\frac{1}{4}$ foot each. With five rulers, that's $5 \times 4 = 20$ parts, so 20 students can receive $\frac{1}{4}$ of a ruler. Thus, the correct answer is option C.

11) Let w be the width of the rectangle. The length is $3w$ and the perimeter is $2w + 2(3w) = 32$. Solving for w, we get:

$$2w + 2(3w) = 32 \Rightarrow 8w = 32 \Rightarrow w = 4,$$

so the length is $3w = 12$, and the area is:

$$w \times 3w = 4 \times 12 = 48.$$

Therefore, the area of rectangle is 48, making option A correct.

12) Initially, the hypotenuse DF can be found using the Pythagorean theorem:

$$DF = \sqrt{DE^2 + EF^2} = \sqrt{8^2 + 6^2} = \sqrt{64 + 36} = \sqrt{100} = 10.$$

After tripling the lengths of DE and EF, the new lengths are 24 and 18, and the new hypotenuse will be 30 (since the sides are scaled by the same factor, the hypotenuse will also be tripled). The new perimeter is:

$$\text{Perimeter} = 24 + 18 + 30 = 72,$$

The new area is:

$$\text{Area} = \frac{1}{2} \times 24 \times 18 = 216,$$

The ratio of the perimeter to the area is:

$$\text{Ratio} = \frac{72}{216} = \frac{1}{3}.$$

The ratio is $\frac{1}{3}$, which is option A.

13) The total of the original seven numbers is $7 \times 24 = 168$. After adding the eighth number, the total becomes $168 + 36 = 204$. The new average is:

$$204 \div 8 = 25.5.$$

Therefore, the new average is 25.5, which should be option D.

14) Let the number of cats be $5x$ and the number of dogs be $8x$. The total number of animals is $5x + 8x = 65$.

8.3 Answers with Explanation

Solving for x gives:
$$5x + 8x = 65 \Rightarrow 13x = 65 \Rightarrow x = 5.$$

Therefore, there are $5x = 5 \times 5 = 25$ cats and $8x = 8 \times 5 = 40$ dogs in the store. To make the ratio 1 : 1, the number of cats and dogs must be equal. Since there are 40 dogs, we need the same number of cats. The store already has 25 cats, so the number of additional cats needed is:

$$40 - 25 = 15.$$

Thus, 15 more cats should be added to achieve a 1 : 1 ratio, which corresponds to option C.

15) The fraction of income saved is:
$$\frac{\$3,000}{\$75,000} = \frac{3}{75} = \frac{1}{25}.$$

Therefore, the correct fractional part of the income saved is $\frac{1}{25}$, which is option C.

16) The inequality is formed by setting the cost of equipment times the number of pieces plus the spent amount less than or equal to the total budget:

$$120x + 18,000 \leq 30,000.$$

This represents the constraint on the number of equipment pieces that can be bought with the remaining budget. Therefore, the option A is correct.

17) To find the minimum score Linda needs on her fifth test, we calculate the total score required for a 75% average across five tests. The total percentage for five tests at 75% each is $5 \times 75\% = 375\%$.

Next, we sum Linda's scores on the first four tests: $70\% + 78\% + 82\% + 75\% = 305\%$.

To find the minimum score needed on the fifth test, we subtract the total of her first four tests from the required total: $375\% - 305\% = 70\%$.

Therefore, Linda must score at least 70% on her fifth test to achieve an average of 75%, making option B the correct answer.

18) To find the growth rate in inches per year, set up a proportion using the given information. The plant grows $2\frac{1}{4}$ inches in $\frac{1}{3}$ of a year. Thus, the proportion is:

$$\frac{2\frac{1}{4} \text{ in}}{\frac{1}{3} \text{ yr}} = \frac{x \text{ in}}{1 \text{ yr}}.$$

Simplifying the fraction $2\frac{1}{4}$ to an improper fraction gives $\frac{9}{4}$. Thus, the equation becomes:

$$\frac{\frac{9}{4} \text{ in}}{\frac{1}{3} \text{ yr}} = \frac{x \text{ in}}{1 \text{ yr}}.$$

Cross-multiplying gives:

$$\frac{1}{3}(x) = \frac{9}{4}(1).$$

Solving for x:

$$x = \frac{9}{4} \div \frac{1}{3} = \left(\frac{9}{4}\right) \times \left(\frac{3}{1}\right) = \frac{27}{4} = 6\frac{3}{4}.$$

Therefore, the plant's growth rate is $6\frac{3}{4}$ inches per year, which is option A.

19) The team paints the fence at a rate of 20 cm per minute. Total length painted in 60 minutes is $20 \frac{cm}{min} \times 60 \, min = 1200 \, cm = 12 \, m$. Since this is $\frac{2}{3}$ of the total length of the fence, we can set up the equation:

$$\frac{2}{3} \times \text{Total Length} = 12 \, m.$$

Solving for the total length, we get:

$$\text{Total Length} = 12 \, m \div \frac{2}{3} = \frac{12 \, m}{1} \times \frac{3}{2} = 18 \, m.$$

Therefore, the total length of the fence is 18 m, which corresponds to choice C.

20) Simplifying the expression:

$$5x^3y^2(3x^3y^2)^2 = 5x^3y^2 \times 9x^6y^4 = 45x^9y^6.$$

The simplified form is $45x^9y^6$, which is option B.

8.3 Answers with Explanation

21) Sarah earns 15% less than Maria, so her hourly rate is:

$$\$40 - 15\% \times \$40 = \$40 - \$40 \times 0.15 = \$40 - \$6 = \$34.$$

Therefore, Sarah earns $34 an hour, making option B correct.

22) Four times last month's visitors is $18,000 \times 4 = 72,000$. With one-fifth changing plans, the attending count is:

$$72,000 - \frac{1}{5} \times 72,000 = 72,000 - 14,400 = 57,600.$$

Thus, 57,600 people are visiting the museum this month, which is option A.

23) The side of the square is the square root of the area, $\sqrt{64} = 8$ inches. The perimeter is 4 times the side length:

$$4 \times 8 \text{ inches} = 32 \text{ inches}.$$

Therefore, the perimeter is 32 inches, which is option D.

24) If we draw a vertical line from E to the segment WZ, we would have four congruent triangle inside the rectangle.

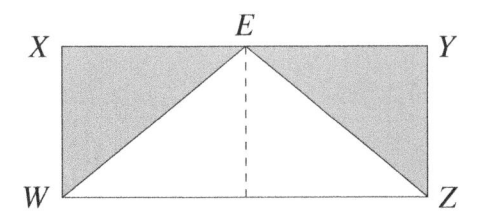

Therefore, the area of the shaded part is half of the area of rectangle:

$$\text{Area of shaded} = \frac{1}{2} \times \text{Area of rectangle} = \frac{1}{2} \times 150 = 75 \text{ square units}.$$

Therefore, the shaded area is 75 square units, option B.

25) First, determine the total number of integers in Set C, which includes even integers from 20 to 200. The number of even numbers in a range can be found by taking half the count of total numbers in that range. For even numbers between 20 and 200, inclusive:

$$\text{Number in Set } C = \frac{200 - 20}{2} + 1 = \frac{180}{2} + 1 = 90 + 1 = 91.$$

Next, identify the numbers that are in both Set C and Set D. These are the even multiples of 5 between 100 and 200, since 200 is the upper limit for Set C. Multiples of 5 that are even are also multiples of 10. Count the number of multiples of 10 between 100 and 200:

$$\text{Number in both Sets } C \text{ and } D = \frac{200 - 100}{10} + 1 = \frac{100}{10} + 1 = 10 + 1 = 11.$$

Finally, calculate the number of integers in Set C that are not in Set D. This is done by subtracting the number of integers that are in both sets from the total number in Set C:

$$\text{Number in Set } C \text{ but not in Set } D = \text{Number in Set } C - \text{Number in both Sets } C \text{ and } D$$
$$= 91 - 11$$
$$= 80.$$

Therefore, there are 80 integers included in Set C but not in Set D, which is option A.

26) Combining like terms:

$$(2x^3 - 3x^2 + 4x) + (7x^3 + 4x^2 - 6x) = 9x^3 + x^2 - 2x.$$

Thus, the simplified form is $9x^3 + x^2 - 2x$, which corresponds to option C.

27) Reflecting the point over the y-axis changes the sign of the x-coordinate:

$$(3, -7) \to (-3, -7).$$

Therefore, the new ordered pair is $(-3, -7)$, and option A is correct.

8.3 Answers with Explanation

28) If the cube of a number is $\frac{8}{27}$, the number is $\sqrt[3]{\frac{8}{27}} = \frac{2}{3}$. Squaring this number:

$$\left(\frac{2}{3}\right)^2 = \frac{4}{9}.$$

Therefore, the square of the number is $\frac{4}{9}$, which corresponds to option A.

29) The designer earns $\$15 \times 8 = \$120$ per day. The expenses are $\$3 \times 8 = \$24$ per day. The net earnings are:

$$\$120 - \$24 = \$96.$$

Thus, the designer makes $\$96$ in one day after expenses, which corresponds to option A.

30) Using the Pythagorean theorem, the shortest distance d from Base Camp B is:

$$d = \sqrt{40^2 + 30^2} = \sqrt{1600 + 900} = \sqrt{2500} = 50 \text{ miles}.$$

Therefore, the shortest distance to Base Camp B is 50 miles, which corresponds to option E.

31) Using the conversion formula:

$$C = \frac{5}{9}(95 - 32) = \frac{5}{9} \times 63 = 35.$$

The equivalent temperature is $35°C$, which corresponds to option B.

32) The total number of fruits is $20 + 15 + 25 + 40 = 100$. The probability of selecting a banana is:

$$\frac{25}{100} = \frac{1}{4}.$$

Therefore, the probability is $\frac{1}{4}$, which corresponds to option A.

33) Observing the pattern, each term increases by one more than the previous increment:

$$4,\ 5(=4+1),\ 7(=5+2),\ 10(=7+3),\ 14(=10+4),\ 19(=14+5),\ 25(=19+6),$$
$$32(=25+7),\ 40(=32+8).$$

The missing term is $25 + 7 = 32$, which corresponds to option E.

34) Let the width be w and the length be $3w$. The perimeter P is given by:

$$P = 2l + 2w = 2(3w) + 2w = 8w.$$

Setting $P = 96$ meters, we find:

$$8w = 96 \Rightarrow w = 12 \text{ meters and } l = 3w = 36 \text{ meters}.$$

Therefore, the length of the park is 36 meters, which corresponds to option D.

35) To find the solution set for x in the inequality $2x + 3 < 19$:

$$2x + 3 < 19 \Rightarrow 2x < 19 - 3 \Rightarrow 2x < 16 \Rightarrow x < 8.$$

The positive integers that satisfy this are $1, 2, 3, 4, 5, 6$, and 7, totaling 7 integers. Thus, option C is correct.

36) Let y be the number of yellow books. We can analyze each choice to determine which cannot be the probability of randomly selecting a yellow book from the shelf:

Total number of books = Red + Blue + Green + Yellow = $15 + 25 + 35 + y$.

A. $\frac{1}{2}$: The probability of choosing a yellow book is $\frac{y}{15+25+35+y} = \frac{y}{75+y} = \frac{1}{2}$. Solving for y gives $2y = 75 + y \Rightarrow y = 75$. This is possible, so option A is not the answer.

B. $\frac{1}{4}$: Similarly, $\frac{y}{75+y} = \frac{1}{4}$. Solving gives $4y = 75 + y \Rightarrow 3y = 75 \Rightarrow y = 25$. This is also possible.

C. $\frac{5}{6}$: Here, $\frac{y}{75+y} = \frac{5}{6}$. Solving gives $6y = 375 + 5y \Rightarrow y = 375$. This is possible.

D. $\frac{1}{5}$: $\frac{y}{75+y} = \frac{1}{5}$. Solving gives $5y = 75 + y \Rightarrow 4y = 75 \Rightarrow y = 18.75$. This is not possible as the number of books cannot be a fraction.

E. $\frac{2}{3}$: Finally, $\frac{y}{75+y} = \frac{2}{3}$. Solving gives $3y = 150 + 2y \Rightarrow y = 150$. This is possible.

Therefore, option D, $\frac{1}{5}$, cannot be the probability of randomly selecting a yellow book from the shelf as it results in a fractional number of books.

8.3 Answers with Explanation

37) To find the slope, solve for y in terms of x:

$$2y = -3x + 6 \Rightarrow y = -\frac{3}{2}x + 3.$$

The slope of the line is the coefficient of x, which is $-\frac{3}{2}$ or -1.5. Therefore, the correct answer is option A, which corresponds to a slope of -1.5.

38) The volume V is given by the product of height, width, and length:

$$V = 4\ cm \times 7\ cm \times 8\ cm = 224\ cm^3.$$

Therefore, the volume of the container is $224\ cm^3$, which corresponds to option C.

39) Subtracting the expressions:

$$\left(6x^3 + 9x^2 - 4x^4\right) - \left(5x^2 - 3x^4 + 7x^3\right) = \left(-4x^4 + 3x^4\right) + \left(6x^3 - 7x^3\right) + \left(9x^2 - 5x^2\right)$$
$$= -x^4 - x^3 + 4x^2.$$

The simplified expression is $-x^4 - x^3 + 4x^2$, which corresponds to option B.

40) In the first year, the sales increase by 15%. So, if the initial sales are represented as 100%, then after a 15% increase, the sales become 115% of the original. This is represented mathematically as $100\% \times (1 + 15\%) = 100\%(1 + 0.15) = 100\%(1.15) = 115\%$.

In the second year, the sales increase by another 25%. However, this increase is applied to the already increased amount (115%), not the original amount. Therefore, the calculation for the second year is $115\% \times (1 + 25\%)$.

Combining these, we get:

$$\text{Total Increase} = 100\% \times (1 + 0.15) \times (1 + 0.25) - 100\%$$
$$= 100\%(1.15 \times 1.25 - 1) = 100\% \times 0.4375 = 43.75\%.$$

This represents a 43.75% increase over the two years, which corresponds to option C.

41) When multiplying exponents with the same base, we add the exponents:

$$(x^4)(x^5) = x^{4+5} = x^9.$$

Therefore, the simplified form is x^9, which corresponds to option B.

42) To maximize one of these six integers, we minimize the others. Since the smallest integer is -12 and all integers are different, so the five numbers are:

$$-12, -11, -10, -9, -8.$$

The sum of these numbers is -50. Therefore, the largest integer must be:

$$-53 - (-50) = -3,$$

which corresponds to option E.

43) Let x be the original quantity. After a 15% increase, the new quantity is $x + 0.15x = 1.15x = 115$ grams. Solving for x gives:

$$x = \frac{115}{1.15} = 100 \text{ grams.}$$

Therefore, the original quantity was 100 grams, option C.

44) Using the proportionality of shadows cast by similar triangles, let h be the height of the flagpole. The ratio of the flagpole's height to the stick's height is equal to the ratio of their shadows:

$$\frac{h}{1.5\,m} = \frac{9\,m}{1\,m} \Rightarrow h = 1.5 \times 9 = 13.5 \text{ meters.}$$

The flagpole is 13.5 meters tall, which is option E.

45) To find the percentage that the tip represents of the total cost, we use the formula:

$$\text{Tip Percentage} = \left(\frac{\text{Tip Amount}}{\text{Total Cost}}\right) \times 100\% = \left(\frac{18}{90}\right) \times 100\% = 20\%.$$

8.3 Answers with Explanation

The tip was 20% of the cost of the meal, option C.

46) Let the number be denoted by *n*. Given that 40% of *n* is 16, we can write:

$$0.40 \times n = 16.$$

Solving for *n* gives:

$$n = \frac{16}{0.40} = 40.$$

Therefore, the number is 40, which is option D.

47) Since *C* is one-third of *D* and *D* is 24, we can find *C* by calculating:

$$C = \frac{1}{3} \cdot 24 = 8.$$

Hence, the value of *C* is 8, which is option A.

48) Sam's speed relative to Lily is $15 - 12 = 3$ kilometers per hour. To catch up 15 kilometers, Sam will take:

$$\text{Time} = \frac{\text{Distance}}{\text{Relative Speed}} = \frac{15}{3} = 5 \text{ hours}.$$

It will take Sam 5 hours to catch up, which is option D.

49) To find the percentage that passed, subtract the number who failed from the total number of students and then divide by the total number of students:

$$\text{Passed} = \frac{60 - 18}{60} \times 100\% = \frac{42}{60} \times 100\% = 70\%.$$

Therefore, 70% of the students passed the exam, which is option B.

50) Let the length of segment *RQ* be *x* cm. Since *PR* is twice as long as *RQ*, $PR = 2x$ cm. The total length of segment *PQ* is the sum of *PR* and *RQ*, which gives us:

$$x + 2x = 60 \; cm \Rightarrow 3x = 60 \; cm \Rightarrow x = \frac{60 \; cm}{3} = 20 \; cm.$$

Therefore, the length of segment *RQ* is 20 *cm*, making option B correct.

9. Practice Test 8

CBEST Math Practice Test

Total number of questions: 50

Total time: 90 Minutes

Calculators are prohibited for the CBEST exam.

9.1 Practices

1) In the following figure, $AE = 2$, $CD = 3$ and $AC = 8$. What is the length of AB?

☐ A. 3.2
☐ B. 4
☐ C. 5.2
☐ D. 3
☐ E. 2.8

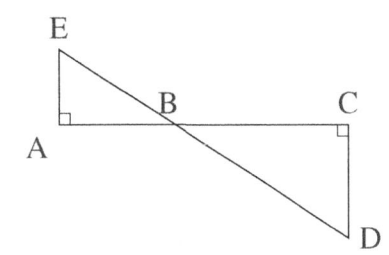

2) An isosceles trapezoid has bases of 10 *cm* and 6 *cm*. If the area of the trapezoid is 16 *cm*^{2}, what is the perimeter of the trapezoid?

- ☐ A. 20 *cm*
- ☐ B. $4(4+\sqrt{2})$ *cm*
- ☐ C. $16+2\sqrt{2}$ *cm*
- ☐ D. $18+2\sqrt{2}$ *cm*
- ☐ E. 36 *cm*

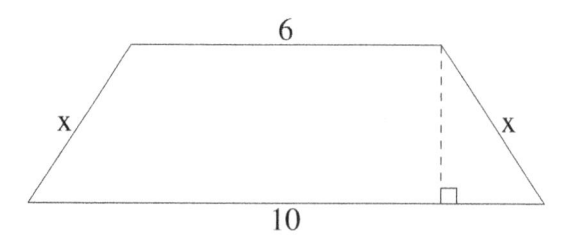

3) If 7 inches on a map represents an actual distance of 200 feet, what actual distance does 21 inches on the map represent?

- ☐ A. 150 feet
- ☐ B. 300 feet
- ☐ C. 600 feet
- ☐ D. 900 feet
- ☐ E. 1200 feet

4) Arrange the following fractions in order from least to greatest:

$$\frac{5}{6}, \frac{7}{8}, \frac{2}{3}, \frac{9}{10}$$

- ☐ A. $\frac{2}{3}, \frac{5}{6}, \frac{7}{8}, \frac{9}{10}$
- ☐ B. $\frac{2}{3}, \frac{7}{8}, \frac{5}{6}, \frac{9}{10}$
- ☐ C. $\frac{7}{8}, \frac{2}{3}, \frac{9}{10}, \frac{5}{6}$
- ☐ D. $\frac{7}{8}, \frac{9}{10}, \frac{2}{3}, \frac{5}{6}$
- ☐ E. $\frac{9}{10}, \frac{7}{8}, \frac{5}{6}, \frac{2}{3}$

5) The graph below represents the scores of eight students in a science quiz. What is the mean (average) score?

9.1 Practices

☐ A. 13.25

☐ B. 15.5

☐ C. 14.25

☐ D. 14.5

☐ E. 15.25

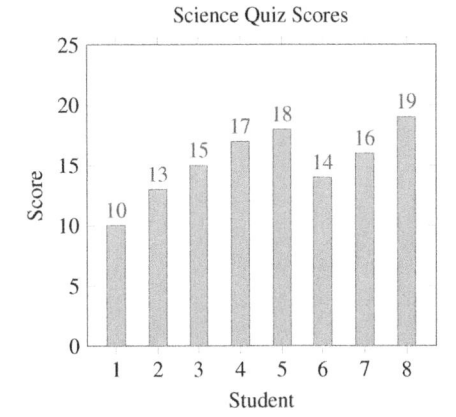

6) Given the populations of men and women in four different cities, E, F, G, and H, with the following diagram (numbers in thousands). what is the maximum ratio (rounding to the nearest hundredth) of women to men?

☐ A. 0.95

☐ B. 0.96

☐ C. 0.97

☐ D. 0.98

☐ E. 0.99

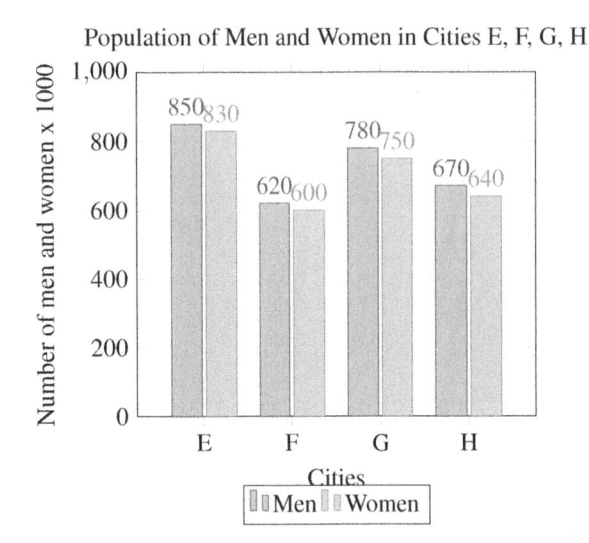

7) Based on the data in question 6, what is the ratio of the percentage (rounded to the nearest hundredth) of men in City E to the percentage of women (rounded to the nearest hundredth) in City G?

☐ A. 0.86

☐ B. 0.91

☐ C. 0.96

☐ D. 1.04

☐ E. 1.06

8) Based on the data in question 6, how many women, in thousands, should be added to city H until the ratio of women to men becomes 1.1?

☐ A. 73

- [ ] B. 75
- [ ] C. 97
- [ ] D. 80
- [ ] E. 82

9) What is the value of 7^3?
   - [ ] A. 7
   - [ ] B. 49
   - [ ] C. 343
   - [ ] D. 2401
   - [ ] E. 16807

10) How many $\frac{1}{4}$ pound hardcover books combined weigh 75 pounds?
    - [ ] A. 75
    - [ ] B. 150
    - [ ] C. 225
    - [ ] D. 300
    - [ ] E. 375

11) Calculate the volume of a square pyramid with a base side length of 9 *m* and a height of 12 *m*.
    - [ ] A. 324 m^3
    - [ ] B. 432 m^3
    - [ ] C. 972 m^3
    - [ ] D. 1080 m^3
    - [ ] E. 1296 m^3

12) The lateral surface area of a cylinder is 200π cm^2. If its height is 8 *cm*, what is the radius of the cylinder?
    - [ ] A. 25 *cm*
    - [ ] B. 20 *cm*
    - [ ] C. 15 *cm*
    - [ ] D. 12.5 *cm*
    - [ ] E. 10 *cm*

9.1 Practices

13) In the figure shown, a circle is inscribed in a square. The area of the circle is 25π. What is the area of the square?

- ☐ A. 25
- ☐ B. 50
- ☐ C. 100
- ☐ D. 125
- ☐ E. 250

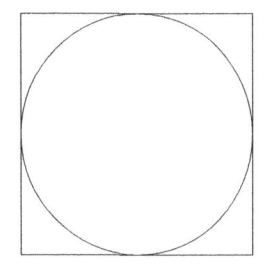

14) List X consists of the numbers $\{2,5,9,11,16\}$, and List Y consists of the numbers $\{7,8,13,18,20\}$. If the two lists are combined, what is the median of the combined list?

- ☐ A. 9
- ☐ B. 10
- ☐ C. 11
- ☐ D. 12
- ☐ E. 13

15) A rectangular garden measuring 20 meters by 14 meters has a rectangular pond in the center measuring 12 meters by 6 meters. What is the area of the garden that is not covered by the pond?

- ☐ A. 208
- ☐ B. 216
- ☐ C. 224
- ☐ D. 280
- ☐ E. 300

16) In triangle DEF, if the measure of angle D is 45 degrees, what is the value of y? (Note: The figure is not drawn to scale)

- ☐ A. 40
- ☐ B. 50
- ☐ C. 55
- ☐ D. 58
- ☐ E. 62

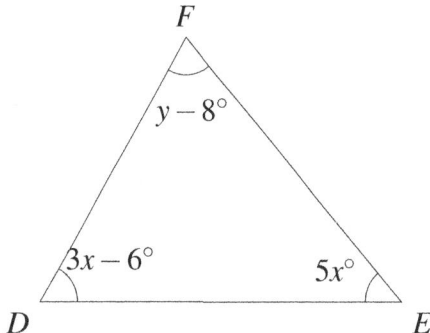

17) Given a dataset on the prevalence of five different diseases across 15 cities, where *a* is the mean (average) of the number of cities experiencing each disease type category, *b* is the mode, and *c* is the median number, which of the following is true?

- A. $a < b < c$
- B. $c < a < b$
- C. $b < c < a$
- D. $a = c$
- E. $a < b = c$

Type of Disease	Number of Cities
A	8
B	6
C	7
D	4
E	4

18) Considering the practice 17, what percent of cities are experiencing the types of diseases A, C, and D respectively? (Round to the nearest integer)

- A. 53%, 60%, 27%
- B. 45%, 30%, 24%
- C. 46%, 55%, 22%
- D. 55%, 45%, 18%
- E. 45%, 75%, 35%

19) Consider the table in practice 17. How many cities must be cleared of disease type C so that the ratio of the number of cities with disease C to the number of cities with disease E becomes 0.5?

- A. 2
- B. 3
- C. 5
- D. 7
- E. 8

20) In a box containing only red and green cards, the probability of randomly selecting a green card is $\frac{2}{5}$. If there are 300 red cards, how many cards are in total in the box?

- A. 150
- B. 300
- C. 450
- D. 500

9.1 Practices

☐ E. 750

21) If $\frac{1}{4a^2} + \frac{1}{8a} = \frac{1}{a^2}$, then a equals?

☐ A. $-\frac{7}{8}$

☐ B. $-\frac{8}{7}$

☐ C. 6

☐ D. 8

☐ E. -9

22) In the figure shown, circle G represents odd numbers greater than 10, circle D represents divisors of 180, and circle M represents multiples of 9. The intersection of all three circles contains a number z. Which of the following numbers could be z?

☐ A. 9

☐ B. 15

☐ C. 18

☐ D. 27

☐ E. 45

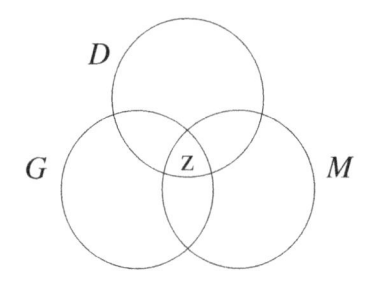

23) Which of the following graphs represents the compound inequality $-7 < -2x + 3 \leq 5$?

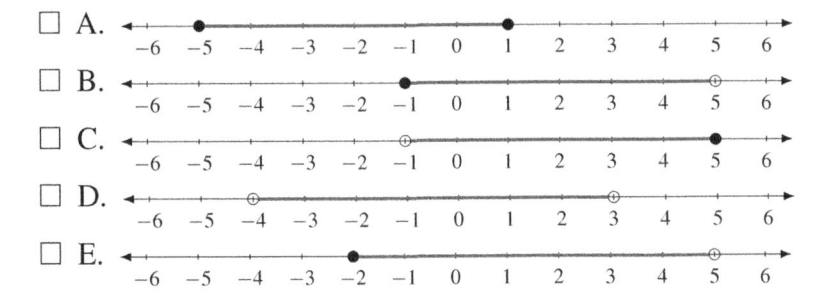

☐ A.
☐ B.
☐ C.
☐ D.
☐ E.

24) A collection of 15 stones has an average weight of 30 g. Among them, the three heaviest stones have an average weight of 50 g each. If these three heaviest stones are removed from the collection, what is the new average weight of the remaining stones?

☐ A. 15 g

☐ B. 25 g

☐ C. 27 g

☐ D. 28 g

☐ E. 30 g

25) During a concert, the ratio of adults to children in the audience is 3 : 4. If there are a total of 21,000 people in the audience, how many adults are there?

- ☐ A. 9,000
- ☐ B. 10,500
- ☐ C. 12,000
- ☐ D. 12,750
- ☐ E. 13,500

26) Find the perimeter of a square playground in meters that has an area of 1,296 m^2.

- ☐ A. 126 m
- ☐ B. 136 m
- ☐ C. 130 m
- ☐ D. 128 m
- ☐ E. 144 m

27) A cake recipe requires $3\frac{1}{4}$ cups of sugar. If you have already $1\frac{1}{2}$ cups, how much more sugar do you need to add?.

- ☐ A. $1\frac{3}{4}$
- ☐ B. $1\frac{1}{2}$
- ☐ C. 2
- ☐ D. $2\frac{1}{4}$
- ☐ E. $2\frac{3}{4}$

28) If $a = \frac{2}{5}$ and $b = \frac{10}{25}$, then what is equal to $\frac{1}{a} \div \frac{b}{2}$?

- ☐ A. 5
- ☐ B. $\frac{1}{2}$
- ☐ C. $\frac{5}{2}$
- ☐ D. $\frac{1}{5}$
- ☐ E. $\frac{25}{2}$

29) Jim's stamp collection will increase to $\frac{4}{3}$ times the current number if he adds 120 stamps. How many stamps does Jim currently have?

9.1 Practices

- ☐ A. 180
- ☐ B. 240
- ☐ C. 360
- ☐ D. 480
- ☐ E. 540

30) The sum of 7 numbers is between 280 and 350. What is the possible range for the average of these numbers?

- ☐ A. 30 to 40
- ☐ B. 40 to 50
- ☐ C. 50 to 60
- ☐ D. 60 to 70
- ☐ E. 70 to 80

31) In the figure below, the size of the angles is displayed with variables. Which is the value of $v+w+z$?

- ☐ A. 140°
- ☐ B. 164°
- ☐ C. 155°
- ☐ D. 172°
- ☐ E. 160°

32) In a pair of similar triangles, the length of a side in the larger triangle is 27 cm, which corresponds to a side of length 12 cm in the smaller triangle. If another side in the smaller triangle measures 8 cm, what is the length of the corresponding side in the larger triangle?

- ☐ A. 16 *cm*
- ☐ B. 18 *cm*
- ☐ C. 20 *cm*
- ☐ D. 22.5 *cm*
- ☐ E. 24 *cm*

33) Oliver is currently twice as old as Julia's age at a specific time in the past. This specific time in the past is when Oliver was exactly as old as Julia is right now. If Julia is currently 12 years old, how old is Oliver?

- ☐ A. 16 years

- [ ] B. 20 years
- [ ] C. 24 years
- [ ] D. 28 years
- [ ] E. 32 years

34) Originally, the length of a garden is 2 meters less than twice its width. If the width of the garden is doubled (without changing the length) resulting in a 10-meter increase in the garden's perimeter, what is the area of the new garden?
- [ ] A. 56 m^2
- [ ] B. 63 m^2
- [ ] C. 72 m^2
- [ ] D. 80 m^2
- [ ] E. 88 m^2

35) Find the value of x in the following diagram.
- [ ] A. 42°
- [ ] B. 44.5°
- [ ] C. 46.5°
- [ ] D. 47°
- [ ] E. 48.5°

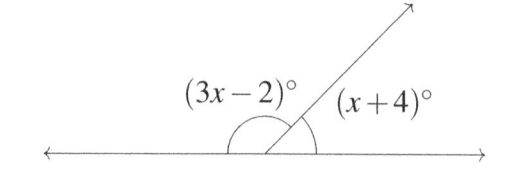

$(3x-2)°$ $(x+4)°$

36) A pie chart represents Dr. Smith's monthly budget distribution. If Dr. Smith spends $550 on groceries, which constitutes 20% of her monthly budget, how much does she spend on utilities that make up 15% of her budget?
- [ ] A. $412.50
- [ ] B. $660.50
- [ ] C. $675.50
- [ ] D. $825.50
- [ ] E. $990.50

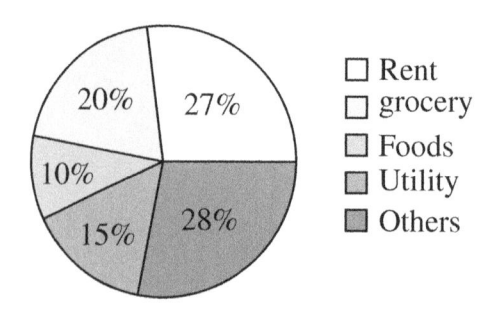

- [ ] Rent
- [ ] grocery
- [ ] Foods
- [ ] Utility
- [ ] Others

37) A circular garden has a walkway around it such that the outer edge of the walkway forms a concentric circle with the garden. If the diameter of the outer edge of the walkway is 10 meters and the diameter of the garden (inner circle) is 6 meters, what is the area of the walkway?

9.1 Practices

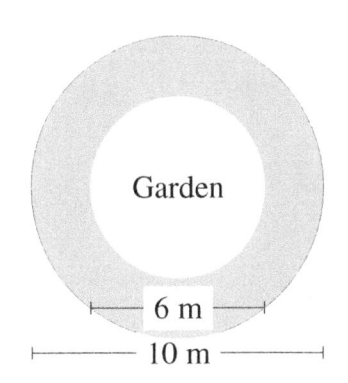

☐ A. $16\pi \ m^2$

☐ B. $19\pi \ m^2$

☐ C. $25\pi \ m^2$

☐ D. $31\pi \ m^2$

☐ E. $44\pi \ m^2$

38) In the rectangle shown below, the area of the rectangle is 48 cm^2, and the perimeter is 28 cm, what are the lengths of the two legs, labeled a and b?

☐ A. $a = 8, \ b = 6$

☐ B. $a = 8, \ b = 3$

☐ C. $a = 6, \ b = 8$

☐ D. $a = 12, \ b = 6$

☐ E. $a = 6, \ b = 4$

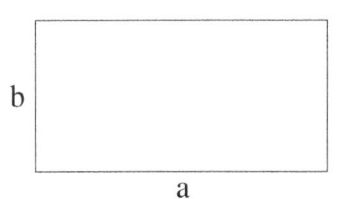

39) In triangle ABC, the external angle at B is $150°$. If C is an obtuse angle, what is the value of angle C?

☐ A. $92°$

☐ B. $96°$

☐ C. $98°$

☐ D. $100°$

☐ E. $118°$

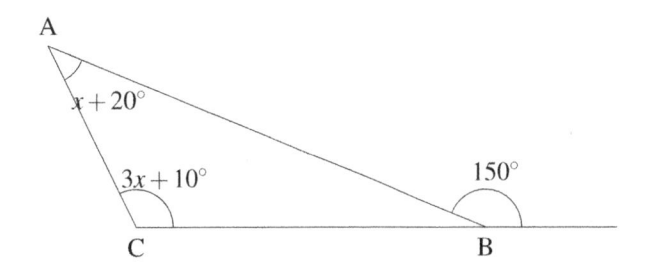

40) If y is inversely proportional to the cube of x, and $y = 3$ when $x = 1$, what is the value of y when $x = 3$?

☐ A. $\frac{1}{9}$

☐ B. $\frac{1}{27}$

☐ C. 9

☐ D. 27

☐ E. 81

41) Emily earns \$540 for the first 36 hours of work in a week, and she receives double her regular hourly rate for any additional hours. She needs \$750 to cover her weekly expenses. How many hours must she work to earn enough money in this week?

☐ A. 39

- ☐ B. 40
- ☐ C. 41
- ☐ D. 42
- ☐ E. 43

42) Calculate 3.75% of 800.

- ☐ A. 10
- ☐ B. 20
- ☐ C. 30
- ☐ D. 40
- ☐ E. 50

43) If z is a real number, and if $z^3 - 27 = 100$, then z lies between which two consecutive integers?

- ☐ A. 3 and 4
- ☐ B. 4 and 5
- ☐ C. 5 and 6
- ☐ D. 6 and 7
- ☐ E. 7 and 8

44) Emily types 72 words per minute. How many words does she type in 25 seconds?

- ☐ A. 21
- ☐ B. 24
- ☐ C. 26
- ☐ D. 30
- ☐ E. 32

45) Which of the following is the scientific notaion of $0.000,000,000,000,000,526,173$?

- ☐ A. 5.26173×10^{17}
- ☐ B. 5.26173×10^{16}
- ☐ C. 526.173×10^{-15}
- ☐ D. 526.173×10^{-18}
- ☐ E. 5.26173×10^{-16}

9.1 Practices

46) Which of the following is the smallest?

- ☐ A. $|-3-1|$
- ☐ B. $|1+3|$
- ☐ C. $|-1-3|$
- ☐ D. $|-1-3|-|3+1|$
- ☐ E. $|1+3|+|-3-1|$

47) A student scores 90% on a test with 50 questions. How many questions did the student answer correctly?

- ☐ A. 40
- ☐ B. 42
- ☐ C. 45
- ☐ D. 48
- ☐ E. 50

48) To finance her car, Olivia borrowed $3,000 at 7% annual interest for 5 years. How much interest will she pay in total?

- ☐ A. $150
- ☐ B. $1,050
- ☐ C. $1,500
- ☐ D. $2,100
- ☐ E. $3,000

49) If m is an odd integer that is less than $\frac{13}{5}$, what is the greatest possible value of m?

- ☐ A. 1
- ☐ B. 2
- ☐ C. 3
- ☐ D. 4
- ☐ E. 5

50) Integer y is divisible by 4. Which of the following expressions is also divisible by 4?

- ☐ A. $y+2$
- ☐ B. $3y+4$

☐ C. $2y+2$
☐ D. $4y+3$
☐ E. $5y+6$

9.2   Answer Keys

1) A. 3.2
2) B. $4(4+\sqrt{2})$ cm
3) C. 600 feet
4) A. $\frac{2}{3}, \frac{5}{6}, \frac{7}{8}, \frac{9}{10}$
5) E. 15.25
6) D. 0.98
7) D. 1.04
8) C. 97
9) C. 343
10) D. 300
11) A. 324 m^3
12) D. 12.5 cm
13) C. 100
14) B. 10
15) A. 208
16) D. 58
17) C. $b < c < a$
18) A. 53%, 60%, 27%
19) D. 7
20) D. 500
21) C. 6
22) E. 45
23) B.
24) B. 25 g
25) A. 9,000

26) E. 144 m
27) A. $1\frac{3}{4}$
28) E. $\frac{25}{2}$
29) C. 360
30) B. 40 to 50
31) D. 172°
32) B. 18 cm
33) A. 16 years
34) D. 80 m^2
35) B. 44.5°
36) A. $412.50
37) A. 16π m^2
38) A. $a = 8, \ b = 6$
39) D. 100°
40) A. $\frac{1}{9}$
41) E. 43
42) C. 30
43) C. 5 and 6
44) D. 30
45) E. 5.26173×10^{-16}
46) D. $|-1-3|-|3+1|$
47) C. 45
48) B. $1,050
49) A. 1
50) B. $3y + 4$

9.3 Answers with Explanation

1) Triangles *BAE* and *BCD* are right-angled triangles and have equal angles. So, these triangles are similar. The ratio of similarity is:

$$k = \frac{AE}{CD} = \frac{2}{3}.$$

Thus,

$$\frac{AB}{BC} = \frac{2}{3} \Rightarrow \frac{AB}{AC} = \frac{2}{5} \Rightarrow \frac{AB}{8} = \frac{2}{5} \Rightarrow AB = \frac{2 \times 8}{5} = 3.2.$$

Thus, the length of *AB* is 3.2 units, making option A correct.

2) To find the perimeter, we need the lengths of all sides. We know the bases are 10 *cm* and 6 *cm*. Let's find the length of the non-parallel sides, labeled as *x*.

The area of an isosceles trapezoid can be found using the formula:

$$\text{Area} = \frac{1}{2}(\text{base}_1 + \text{base}_2) \times \text{height}.$$

Given that the area is 16 cm^2:

$$16 = \frac{1}{2}(10+6) \times h \Rightarrow 16 = 8h \Rightarrow h = 2 \ cm$$

Now, we apply the Pythagorean theorem to find *x*. Each leg of the trapezoid forms a right triangle with the height and half the difference of the bases:

$$x^2 = h^2 + \left(\frac{10-6}{2}\right)^2 \Rightarrow x^2 = 2^2 + 2^2 \Rightarrow x^2 = 4+4 \Rightarrow x = \sqrt{8} = 2\sqrt{2} \ cm$$

Finally, the perimeter is the sum of all sides:

$$\text{Perimeter} = 10 + 6 + 2x = 16 + 4\sqrt{2} = 4(4 + \sqrt{2}) \ cm$$

Therefore, the correct answer is option B.

9.3 Answers with Explanation

3) Set up a proportion based on the map scale:

$$\frac{7 \text{ inches}}{200 \text{ feet}} = \frac{21 \text{ inches}}{x \text{ feet}} \Rightarrow x = \frac{21 \text{ inches} \times 200 \text{ feet}}{7 \text{ inches}} = 600 \text{ feet}.$$

Therefore, 21 inches on the map represent 600 feet, making option C correct.

4) To arrange the fractions in order from least to greatest, we find a common denominator and compare the equivalent fractions.

The least common denominator for 3, 6, 8, and 10 is 120. We convert each fraction to have this common denominator:

$$\frac{5}{6} = \frac{100}{120}, \quad \frac{7}{8} = \frac{105}{120}, \quad \frac{2}{3} = \frac{80}{120}, \quad \frac{9}{10} = \frac{108}{120}.$$

With these equivalent fractions, it becomes easier to compare them.

We find that $\frac{80}{120}$ (equivalent to $\frac{2}{3}$) is the smallest, followed by $\frac{100}{120}$ (equivalent to $\frac{5}{6}$), $\frac{105}{120}$ (equivalent to $\frac{7}{8}$), and $\frac{108}{120}$ (equivalent to $\frac{9}{10}$) being the largest.

Therefore, the correct order from least to greatest is option A: $\frac{2}{3}, \frac{5}{6}, \frac{7}{8}, \frac{9}{10}$.

5) To find the mean score, add all the scores together and then divide by the number of scores:

$$\text{Mean} = \frac{10+13+15+17+18+14+16+19}{8} = \frac{122}{8} = 15.25.$$

The mean score is 15.25, making option E correct.

6) Calculate the ratio of women to men for each city and identify the maximum value:

$$\text{City E: } \frac{830}{850} \approx 0.98, \quad \text{City F: } \frac{600}{620} \approx 0.97, \quad \text{City G: } \frac{750}{780} \approx 0.96, \quad \text{City H: } \frac{640}{670} \approx 0.96.$$

The maximum ratio is approximately 0.98 in City E. Therefore, the correct answer is option D.

7) Calculate the percentages:

$$\text{Percentage of Men in City E: } \frac{850}{\text{Total in E}} = \frac{850}{1680} \approx 0.51,$$

$$\text{Percentage of Women in City G: } \frac{750}{\text{Total in G}} = \frac{750}{1530} \approx 0.49.$$

The ratio is approximately $\frac{0.51}{0.49} \approx 1.04$, making option D correct.

8) Let x be the number of women added to city H to achieve the ratio of 1.1:

$$\frac{640+x}{670} = 1.1.$$

Solve for x:

$$640 + x = 1.1 \times 670 \Rightarrow x = 737 - 640 = 97.$$

Therefore, 97 thousands women should be added to city H, making option C correct.

9) The value of 7^3 is calculated as:

$$7 \times 7 \times 7 = 343.$$

Therefore, the value of 7^3 is 343, making option C correct.

10) To find out how many $\frac{1}{4}$ pound books make up 75 pounds, divide the total weight by the weight of one book:

$$\frac{75}{\frac{1}{4}} = 75 \times 4 = 300.$$

Therefore, 300, $\frac{1}{4}$ pound books weigh 75 pounds, making option D correct.

11) The volume V of a square pyramid is given by $V = \frac{1}{3} \cdot$ (base area) $\cdot$ (height):

$$V = \frac{1}{3} \cdot 9^2 \cdot 12 = \frac{1}{3} \cdot 81 \cdot 12 = 324 \ m^3.$$

Therefore, the volume of the pyramid is 324 m^3, making option A correct.

12) The lateral surface area A of a cylinder is given by $A = 2\pi r h$ where r is the radius and h is the height:

$$200\pi = 2\pi r \times 8 \Rightarrow r = \frac{200\pi}{2\pi \times 8} = \frac{200}{16} = 12.5 \ cm.$$

Therefore, the radius of the cylinder is 12.5 cm, making option D correct.

9.3 Answers with Explanation

13) The area of a circle is $A_{circle} = \pi r^2$. Given that $A_{circle} = 25\pi$, we can find the radius r as follows:

$$r^2 = 25 \Rightarrow r = 5.$$

The diameter of the circle is equal to the side length of the square, so the side length s of the square is $2r = 10$. The area of the square is:

$$A_{square} = s^2 = 10^2 = 100.$$

Thus, the area of the square is 100 square units, making option C correct.

14) The combined list in ascending order is $\{2, 5, 7, 8, 9, 11, 13, 16, 18, 20\}$. The median is the middle number of the ordered list. With 10 numbers, the median will be the average of the 5th and 6th numbers:

$$\text{Median} = \frac{9+11}{2} = \frac{20}{2} = 10.$$

Thus, the median of the combined list is 10, maikng option B correct.

15) To find the area of the non-pond part, subtract the area of the pond from the area of the garden:

$$\text{Area}_{garden} = 20 \times 14 = 280 \ m^2,$$

$$\text{Area}_{pond} = 12 \times 6 = 72 \ m^2,$$

$$\text{Area}_{non\text{-}pond} = \text{Area}_{garden} - \text{Area}_{pond} = 280 - 72 = 208 \ m^2.$$

Therefore, the area of the garden not covered by the pond is $208 \ m^2$, which is option A.

16) The angle D is 45 degrees:

$$3x - 6 = 45 \Rightarrow 3x = 51 \Rightarrow x = 17.$$

Since the sum of angles in any triangle is $180°$, set up the equation:

$$3x - 6 + 5x + y - 8 = 180 \Rightarrow 45 + 5(17) + y - 8 = 180 \Rightarrow y = 58.$$

Therefore, the value of y is 58, making option D correct.

17) Calculate the mean a, mode b, and median c from the given data:

$$a = \frac{\text{Sum of all cities}}{5}, \quad b = \text{Most frequent city count}, \quad c = \text{Middle city count when ordered}.$$

We get:

$$a = \frac{8+5+9+4+4}{5} = 6, \quad b = 4, \quad c = 5.$$

Therefore, option C is correct.

18) To find the percentage of cities affected by each disease, we use the formula:

$$\text{Percentage} = \left(\frac{\text{Number of cities affected by disease}}{15}\right) \times 100\%.$$

Thus, we get:

$$\text{disease A} = \frac{8}{15} \times 100\% \approx 53.33\%, \quad \text{disease C} = \frac{9}{15} \times 100\% = 60\%,$$
$$\text{disease D} = \frac{4}{15} \times 100\% \approx 26.66\%.$$

Therefore, the correct answer is option A.

19) From the table, initially, there are 9 cities with disease C and 4 cities with disease E. Let the number of cities to be cleared of disease C be x. To achieve the desired ratio:

$$\frac{9-x}{4} = 0.5 \Rightarrow 9-x = 4 \times 0.5 \Rightarrow 9-x = 2 \Rightarrow x = 7.$$

Therefore, 7 cities must be cleared of disease C, making option D correct.

20) The probability of randomly selecting a red card is: $1 - \frac{2}{5} = \frac{3}{5}$. There are 300 red cards:

$$\frac{300}{T} = \frac{3}{5} \Rightarrow T = \frac{300 \times 5}{3} = 500.$$

The total number of cards is 500, making option D correct.

9.3 Answers with Explanation

21) Starting with the given equation and choosing the common denominator $8a^2$:

$$\frac{1}{4a^2} + \frac{1}{8a} = \frac{1}{a^2} \Rightarrow \frac{2}{8a^2} + \frac{a}{8a^2} = \frac{8}{8a^2} \Rightarrow 2 + a = 8 \Rightarrow a = 6.$$

Therfore, the correct answer is option C.

22) Circle G represents odd numbers greater than 10, circle D represents divisors of 180, and circle M represents multiples of 9. The intersection of all three circles means z must satisfy all three conditions.

Let's check each option:

- A. 9: This is not less than 10, hence not in G.
- B. 15: This is an odd number greater than 10 and a divisor of 180, but it is not a multiple of 9.
- C. 18: This is not an odd number, hence not in G.
- D. 27: This is an odd number greater than 10 and a multiple of 9, but it is not a divisor of 180.
- E. 45: This is an odd number greater than 10, a divisor of 180 (as $180 = 45 \times 4$), and a multiple of 9 (as $45 = 9 \times 5$).

Therefore, the number that fits all three conditions is 45, which is option E.

23) First, subtract 3 from each part of the inequality:

$$-7 - 3 < -2x + 3 - 3 \leq 5 - 3 \Rightarrow -10 < -2x \leq 2.$$

Now, divide each part by -2. Remember that dividing or multiplying by a negative number reverses the inequality signs:

$$\frac{-10}{-2} > \frac{-2x}{-2} \geq \frac{2}{-2} \Rightarrow 5 > x \geq -1.$$

So, the solution of the compound inequality is $-1 \leq x < 5$. This means x can take any value greater than or equal to -1 but less than 5. Therefore, the correct option is B.

24) Initially, the total weight is 15×30 g $= 450$ g. The total weight of the three heaviest stones is 3×50 g $= 150$ g. Removing these, we have 450 g $- 150$ g $= 300$ g left. The average weight of the remaining $15 - 3 = 12$ stones is:

$$\text{New average} = \frac{300 \text{ g}}{12} = 25 \text{ g}.$$

Thus, the new average weight is 25 g, making option B correct.

25) The total ratio parts are $3 + 4 = 7$. Each part represents $\frac{21,000}{7} = 3,000$ people. Hence, the number of adults, which is 3 parts, is:

$$\text{Adults} = 3 \times 3,000 = 9,000.$$

Therefore, there are 9,000 adults in the audience, which makes option A correct.

26) The side of the square is the square root of the area:

$$\text{Side} = \sqrt{1,296 \, m^2} = 36 \, m.$$

The perimeter is four times the side length:

$$\text{Perimeter} = 4 \times 36 \, m = 144 \, m.$$

Hence, the perimeter of the playground is 144 m, making option E the correct choice.

27) The additional sugar needed is the difference between the required amount and what you already have:

$$3\frac{1}{4} - 1\frac{1}{2} = \frac{13}{4} - \frac{6}{4} = \frac{7}{4} = 1\frac{3}{4}.$$

Therefore, you need to add $1\frac{3}{4}$ more cups of sugar, making option A correct.

28) First, calculate the reciprocal of a and the value of b divided by 2:

$$\frac{1}{a} = \frac{1}{\frac{2}{5}} = \frac{5}{2}, \text{ and } \frac{b}{2} = \frac{\frac{10}{25}}{2} = \frac{10}{50} = \frac{1}{5}.$$

Now divide $\frac{5}{2}$ by $\frac{1}{5}$:

$$\frac{\frac{5}{2}}{\frac{1}{5}} = \frac{5}{2} \times \frac{5}{1} = \frac{25}{2}.$$

The result of the division is $\frac{25}{2}$, which corresponds to option E.

9.3 Answers with Explanation

29) Let n be the current number of stamps. Then $n + 120 = \frac{4}{3}n$. Solving for n gives:

$$n + 120 = \frac{4}{3}n \Rightarrow \frac{4}{3}n - n = 120 \Rightarrow \frac{1}{3}n = 120 \Rightarrow n = 360.$$

Therefore, Jim currently has 360 stamps, making option C correct.

30) The average A of the 7 numbers, given that their sum S is between 280 and 350, is $A = \frac{S}{7}$. Therefore:

$$280 < S < 350 \Rightarrow \frac{280}{7} < \frac{S}{7} < \frac{350}{7} \Rightarrow 40 < A < 50.$$

Therefore, the range for the average is from 40 to 50, making option B correct.

31) Two angles v and w are vertically opposite angles. Thus $v = w$. On the other hand, angle v is an adjacent supplementary angle with both $2v + 6$ and $2z + 10$. Thus, we have:

$$v + 2v + 6 = 180 \Rightarrow 3v = 174 \Rightarrow v = 58.$$

$$v + 2z + 10 = 180 \Rightarrow 58 + 2z = 170 \Rightarrow 2z = 112 \Rightarrow z = 56.$$

Thus, we get:

$$v + w + z = 58 + 58 + 56 = 172.$$

Therefore, the correct answer is opiton D.

32) The sides of similar triangles are proportional. The ratio of the sides of the smaller triangle to the larger triangle is $\frac{12}{27}$. To find the corresponding side in the larger triangle, we set up the proportion $\frac{8}{x} = \frac{12}{27}$, where x is the corresponding side length in the larger triangle. Solving for x gives:

$$x = \frac{8 \times 27}{12} = \frac{216}{12} = 18.$$

Therefore, the corresponding side length in the larger triangle is 18 *cm*, which makes option B correct.

33) Let's denote Oliver's current age as O and Julia's current age as $J = 12$. Let x be the number of years ago when Oliver was as old as Julia is now. Therefore, we have:

$$O - x = J \Rightarrow O - x = 12.$$

At that time, Julia was $J - x = 12 - x$ years old. Since Oliver is now twice as old as Julia was at that time, we have:

$$O = 2(12 - x).$$

Combining the two equations, we get:

$$O - x = 12 \text{ and } O + 2x = 24.$$

Solving this system of equations, we find that $x = 4$ and $O = 16$. Thus, Oliver is 16 years old, making option A correct.

34) Let the original width of the garden be w meters. Therefore, the original length l is $2w - 2$ meters. The original perimeter P is:

$$P = 2l + 2w = 2(2w - 2) + 2w = 4w - 4 + 2w = 6w - 4.$$

When the width is doubled, the new width is $2w$ meters. The length remains $2w - 2$ meters. The new perimeter P' is:

$$P' = 2(2w - 2) + 2(2w) = 4w - 4 + 4w = 8w - 4.$$

Given that the increase in perimeter is 10 meters, we have $P' - P = 10$, so:

$$8w - 4 - (6w - 4) = 10 \Rightarrow w = 5.$$

The new area A is:

$$A = length \times width = (2w - 2) \times 2w = (2 \times 5 - 2) \times 2 \times 5 = 8 \times 10 = 80 \ m^2.$$

Therefore, the area of the new garden is 80 square meters, making option D correct.

9.3 Answers with Explanation

35) Since the angles are supplementary, their sum is 180°. We set up the equation as follows:

$$(x+4) + (3x-2) = 180$$

Simplifying, we find x:

$$4x + 2 = 180 \Rightarrow 4x = 178 \Rightarrow x = 44.5°.$$

Therefore, the value of x is 44.5°, making option B correct.

36) First, calculate the total monthly budget using the cost of groceries:

$$\text{Total Budget} = \frac{\$550}{0.20} = \$2750.$$

Then, calculate the amount spent on utilities:

$$\text{Utilities} = \$2750 \times 0.15 = \$412.50.$$

Thus, Dr. Smith spends $412.50 on utilities, making option A correct.

37) First, calculate the area of the larger circle (walkway + garden):

$$A_{\text{larger}} = \pi r_{\text{larger}}^2 = \pi \left(\frac{10}{2}\right)^2 = 25\pi \ m^2.$$

Then, calculate the area of the smaller circle (garden):

$$A_{\text{smaller}} = \pi r_{\text{smaller}}^2 = \pi \left(\frac{6}{2}\right)^2 = 9\pi \ m^2.$$

The area of the walkway is the difference between the two areas:

$$A_{\text{walkway}} = A_{\text{larger}} - A_{\text{smaller}} = 25\pi - 9\pi = 16\pi \ m^2.$$

Therefore, the area of the walkway is $16\pi \ m^2$, making option A correct.

38) The area of rectangle is given by:

$$\text{Area} = ab = 48.$$

The perimeter P is the sum of the sides:

$$P = 2(a+b) = 28 \ cm.$$

We have two equations $ab = 48$ and $a+b = 14$, so we can find a and b as follows:

$$ab = 48 \Rightarrow b = \frac{48}{a}.$$

$$a+b = 14 \Rightarrow a + \frac{48}{a} = 14 \Rightarrow a^2 + 48 = 14a \Rightarrow (a-8)(a-6) = 0 \Rightarrow a = 8 \text{ or } a = 6.$$

From $ab = 48$, we get: $b = 6$ or $b = 8$. Since $a > b$, the correct option is A.

39) Since the external angle at B is $150°$, the internal angle at B would be $180° - 150° = 30°$. Using the fact that the sum of angles in a triangle is $180°$, we have:

$$30 + 3x + 10 + x + 20 = 180.$$

Solving for x, we get: $x = 30°$. Thus, the angle C is:

$$3x + 10 = 3 \times 30 + 10 = 100°,$$

which is option D.

40) Since y is inversely proportional to x^3, we have $y = \frac{k}{x^3}$ or $y \cdot x^3 = k$ for some constant k. When $x = 1$, $y = 3$, so $k = 3 \cdot 1^3 = 3$. When $x = 3$, we have:

$$y = \frac{k}{x^3} = \frac{3}{3^3} = \frac{1}{9}.$$

Thus, $y = \frac{1}{9}$ when $x = 3$, making option A correct.

41) Emily's regular hourly rate is $\frac{\$540}{36 \text{ hours}} = \$15$ per hour. Let h be the additional hours she needs to work.

9.3 Answers with Explanation

Her total earnings for additional hours at double rate are $2 \times \$15 \times h$. She needs $\$750$. So:

$$\$540 + (2 \times \$15 \times h) = \$750.$$

Solving for h gives:

$$30h = \$750 - \$540 = \$210 \Rightarrow h = \frac{\$210}{\$30} = 7.$$

Therefore, she must work at least 7 additional hours, making her total $36 + 7 = 43$ hours, which is option E.

42) We have:

$$3.75\% \text{ of } 800 = \frac{3.75}{100} \times 800 = 30.$$

Therefore, 3.75% of 800 is 30, making option C correct.

43) Rearrange the equation:

$$z^3 = 100 + 27 = 127.$$

Since $5^3 = 125$ and $6^3 = 216$, z must lie between 5 and 6, making option C correct.

44) Every minute is 60 seconds. So, we have:

$$\text{Words in 25 seconds} = 72 \text{ words/minute} \times \frac{25}{60} \text{ minutes} = 30 \text{ words}.$$

Therefore, Emily types 30 words in 25 seconds, making option D correct.

45) To convert a number into scientific notation, you move the decimal point to the right or left until only one non-zero digit remains on the left. The number of places you move the decimal point determines the exponent on 10. If you move the decimal to the right, the exponent is negative, and if you move it to the left, the exponent is positive. In this case, the decimal point is moved 16 places to the right, making the exponent -16.

$$0.000,000,000,000,000,526,173 = 5.26173 \times 10^{-16}.$$

Thus, the correct scientific notation is 5.26173×10^{-16}, making option E correct.

46) To determine the smallest value, we evaluate each absolute value expression and perform the indicated operations:

A. $|-3-1| = |-4| = 4.$

B. $|1+3| = |4| = 4.$

C. $|-1-3| = |-4| = 4.$

D. $|-1-3| - |3+1| = |-4| - |4| = 4 - 4 = 0.$

E. $|1+3| + |-3-1| = |4| + |-4| = 4 + 4 = 8.$

After calculating, we see that the expression in option D gives us the smallest value, which is 0.

47) The correct answers is:

$$90\% \text{ of } 50 \text{ questions} = \frac{90}{100} \times 50 = 45 \text{ questions.}$$

Therefore, the student answered 45 questions correctly, making option C the correct answer.

48) Considering the formula for calculating Interest, we have:

$$\text{Interest} = \text{Principal} \times \text{Rate} \times \text{Time} = \$3,000 \times 0.07 \times 5 = \$1,050.$$

Thus, Olivia will pay $\$1,050$ interest, at the end of 5 years, making option B the correct answer.

49) By converting the fraction to a decimal number, we have:

$$\frac{13}{5} = 2.6.$$

Since m is an odd integer less than 2.6, the greatest possible odd integer is 1, making option A correct.

50) To determine if an expression is divisible by 4, each term within the expression must be divisible by 4. Since y is given as divisible by 4, we can write y as $4k$, where k is an integer.

Substituting $4k$ into the expressions:

- For A: $y + 2 = 4k + 2$, which is not divisible by 4 as 2 is not a multiple of 4.
- For B: $3y + 4 = 3(4k) + 4 = 12k + 4$, and since $12k$ and 4 are both multiples of 4, their sum is also divisible by 4.
- For C: $2y + 2 = 2(4k) + 2 = 8k + 2$, which is not divisible by 4 due to the addition of 2.

- For D: $4y+3 = 4(4k)+3 = 16k+3$, which is not divisible by 4 as 3 is not a multiple of 4.
- For E: $5y+6 = 5(4k)+6 = 20k+6$, and since 6 is not a multiple of 4, the expression is not divisible by 4.

Therefore, the expression $3y+4$ is the only one that remains divisible by 4 regardless of the value of k, making option B correct.

10. Practice Test 9

CBEST Math Practice Test

Total number of questions: 50

Total time: 90 Minutes

Calculators are prohibited for the CBEST exam.

10.1 Practices

1) The capacity of a green container is 25% larger than the capacity of a yellow container. If the green container can hold 40 equal-sized balls, how many of the same balls can the yellow container hold?

☐ A. 28
☐ B. 30
☐ C. 32
☐ D. 34

10.1 Practices

☐ E. 36

2) Laura spent $45 on a dress. This was $15 less than double what she spent on a pair of shoes. How much were the shoes?

☐ A. $20
☐ B. $25
☐ C. $30
☐ D. $35
☐ E. $40

3) What is the greatest integer less than $-\frac{45}{6}$?

☐ A. −7
☐ B. −8
☐ C. −6
☐ D. −9
☐ E. −10

4) The measures of the angles of a quadrilateral are in the ratio $1:2:3:4$. What is the measure of the largest angle?

☐ A. 72°
☐ B. 108°
☐ C. 144°
☐ D. 180°
☐ E. 216°

5) In the following figure, line C is parallel to line D, and there are two angles related to x. What is the value of x?

☐ A. 20
☐ B. 25
☐ C. 30
☐ D. 35
☐ E. 40

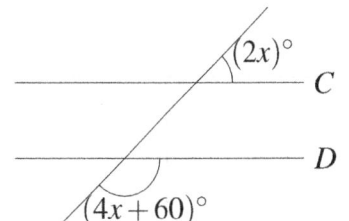

6) The mean of 60 test scores was calculated as 90. However, it turned out that one of the scores was misread as 98 when it was actually 74. What is the corrected mean?

☐ A. 90.5
☐ B. 90.6
☐ C. 89.8
☐ D. 89.7
☐ E. 89.6

7) Solve the compound inequality $-6 \leq 5x - 10 < 20$.

☐ A. $0.8 \leq x < 6$
☐ B. $0.8 < x \leq 6$
☐ C. $1 \leq x < 5$
☐ D. $1 < x \leq 6$
☐ E. $2 \leq x \leq 5$

8) In the following figure, if $a = 2b$, find area of the shaded trapezoid inside the rectangle.

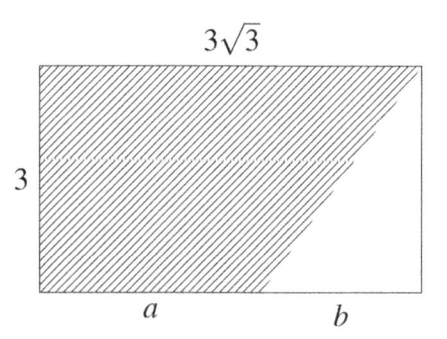

☐ A. $7.5\sqrt{3}$
☐ B. $8\sqrt{3}$
☐ C. $8.5\sqrt{3}$
☐ D. $9\sqrt{3}$
☐ E. $9.5\sqrt{3}$

9) Which graph shows a non-proportional linear relationship between x and y?

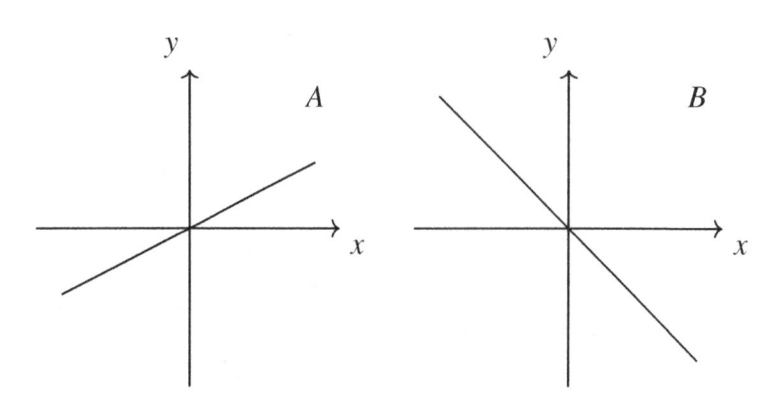

10.1 Practices   235

C

D

E

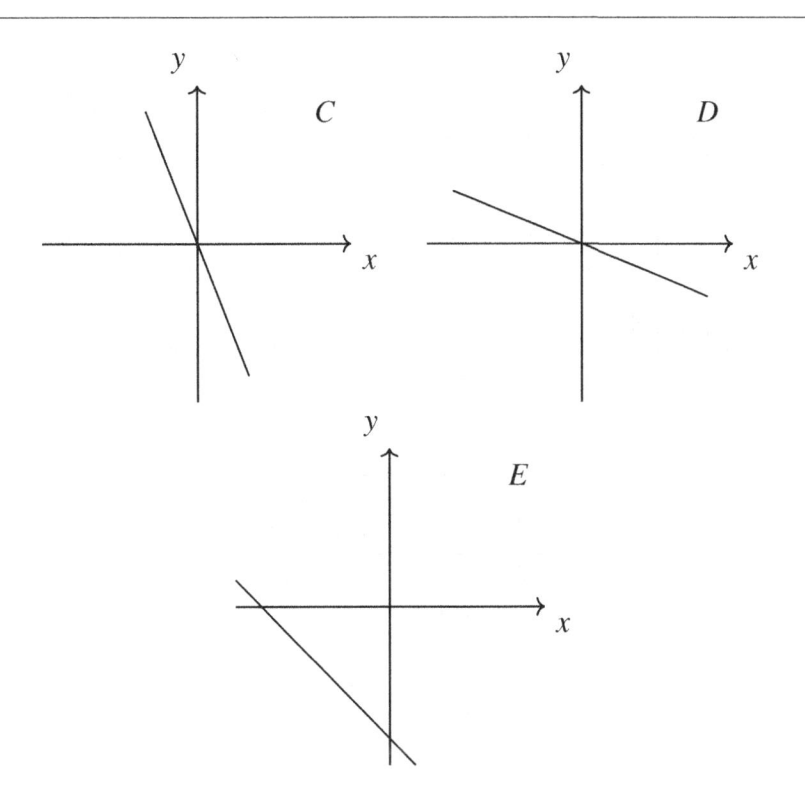

☐ A. Graph A

☐ B. Graph B

☐ C. Graph C

☐ D. Graph D

☐ E. Graph E

10) In the following figure, what is the value of *x*?

☐ A. 16 cm

☐ B. 18 cm

☐ C. 10 cm

☐ D. 14 cm

☐ E. 12 cm

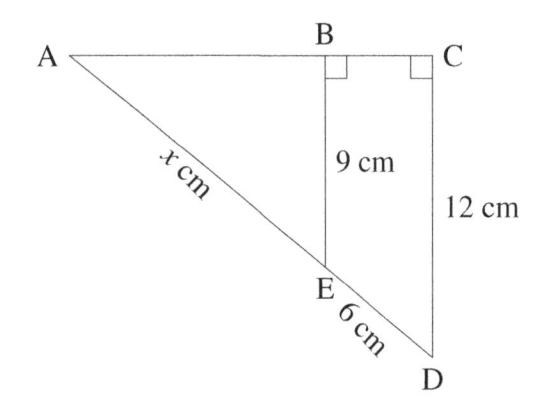

11) Carlos opened an account with a deposit of $4,500. This account earns 4% simple interest annually. How many years will it take him to earn $720 on his $4,500 deposit?

☐ A. 3

☐ B. 4

☐ C. 5

☐ D. 6

☐ E. 7

12) Emily picked $2\frac{3}{5}$ kilograms of oranges, and John picked $3\frac{3}{4}$ kilograms of oranges. How many kilograms in total did they pick?

☐ A. 5.25

☐ B. 5.75

☐ C. 6

☐ D. 6.25

☐ E. 6.35

13) A list of consecutive even integers starts with $2m$ and ends with $2n$. If $2n - 2m = 92$, how many integers are in the list?

☐ A. 23

☐ B. 38

☐ C. 46

☐ D. 47

☐ E. 47

14) A rope that is $4\frac{3}{5}$ meters long is cut into 3 pieces of different lengths. The shortest piece has a length of y meters. Which inequality expresses all possible values of y?

☐ A. $y < 1\frac{1}{2}$

☐ B. $y > 1\frac{1}{2}$

☐ C. $y < 1\frac{8}{15}$

☐ D. $y > 1\frac{3}{4}$

☐ E. $y < 1\frac{3}{4}$

15) In a school, course grades range from 0 to 100. Ben took 7 courses and his mean course grade was 84. Jessica took 10 courses. If both students have the same sum of course grades, what was Jessica's mean grade?

☐ A. 50

10.1 Practices

☐ B. 58.8

☐ C. 70

☐ D. 80

☐ E. 85

16) The sum of three numbers a, b, and c is 105. The ratio of a to b is $3:4$, and the ratio of b to c is $4:7$. What is the value of b?

☐ A. 12

☐ B. 15

☐ C. 24

☐ D. 28

☐ E. 30

17) Provide a number line representation for the solution of the inequality $-4 < -2x \leq 6$.

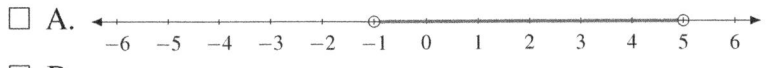

☐ A.

☐ B.

☐ C.

☐ D.

☐ E.

18) The set T consists of all odd integers greater than 10 and less than 50. What is the mean of the numbers in T?

☐ A. 20

☐ B. 25

☐ C. 29

☐ D. 30

☐ E. 35

19) If $\sqrt{3z} = \sqrt{7w}$, what is the ratio of w to z?

☐ A. $\frac{3}{7}$

☐ B. $\frac{7}{3}$

☐ C. $\frac{1}{2}$

☐ D. 2

☐ E. $\frac{1}{\sqrt{2}}$

20) A square *ABCD* has a side length of 6 meters. A line connects the midpoint *M* of *AB* with the midpoint *N* of *BC*. Also, a line connects *N* with *D*. Calculate the area of the polygon *AMND* formed inside the square.

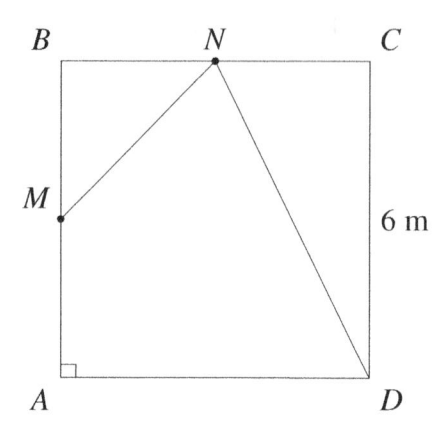

☐ A. $9 \, m^2$

☐ B. $18 \, m^2$

☐ C. $22.5 \, m^2$

☐ D. $24.5 \, m^2$

☐ E. $26.5 \, m^2$

21) In a group of 60 students, 70% cannot swim. How many students can swim?

☐ A. 15

☐ B. 18

☐ C. 20

☐ D. 22

☐ E. 24

22) $10,000 is distributed equally among 16 people. How much money will each person get?

☐ A. $500

☐ B. $600

☐ C. $625

☐ D. $750

☐ E. $800

23) A box contains 8 green sticks, 6 blue sticks, and 4 yellow sticks. Emma picks one without looking. What is the probability that the stick will be green?

☐ A. $\frac{1}{3}$

☐ B. $\frac{2}{5}$

☐ C. $\frac{1}{2}$

10.1 Practices

☐ D. $\frac{4}{9}$

☐ E. $\frac{6}{18}$

24) In the figure below, a regular hexagon is inscribed in a circle. Calculate the shaded area, knowing that the radius of the circle is 4 cm ($\pi \approx 3.14$, $\sqrt{3} \approx 1.73$).

☐ A. 12.68 cm²

☐ B. 10.46 cm²

☐ C. 9.52 cm²

☐ D. 12.26 cm²

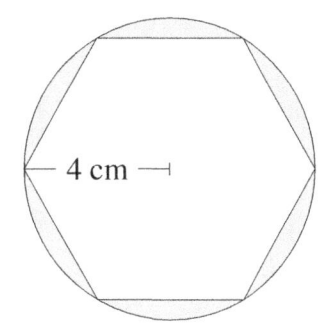

☐ E. 8.72 cm²

25) The price of a commodity was raised from $7.50 to $8.00. What was nearest value to the percent increase in the price?

☐ A. 5%

☐ B. 6%

☐ C. 6.67%

☐ D. 7%

☐ E. 8%

26) In a bag of red and green marbles, the ratio of red marbles to green marbles is 5 : 6. If the bag contains 180 green marbles, how many red marbles are there?

☐ A. 120

☐ B. 150

☐ C. 200

☐ D. 250

☐ E. 300

27) If $\frac{7}{9}$ of a number is 126, find the number.

☐ A. 162

☐ B. 252

☐ C. 324

☐ D. 567

☐ E. 810

28) A smoothie mixture contains $\frac{2}{5}$ of a cup of banana puree and $\frac{2}{25}$ of a cup of strawberry puree. How many cups of banana puree per cup of strawberry puree does the mixture contain?

☐ A. $\frac{25}{2}$

☐ B. 10

☐ C. 2

☐ D. $\frac{5}{2}$

☐ E. 5

29) In the figure below, *ABCD* is a square inscribed in a circle with center *O*. If the side of the square is 4 cm, calculate the area, in square centimeters, of the shaded region outside the square but inside the circle ($\pi \approx 3.14$).

☐ A. 3.14

☐ B. 6.28

☐ C. 8.56

☐ D. 9.12

☐ E. 10.24

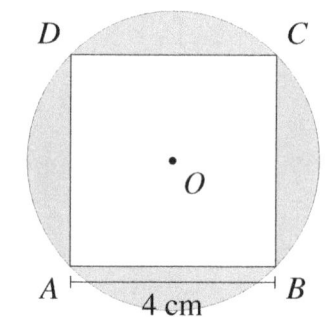

30) The set of possible values of *p* is $\{6,9,12\}$. What is the set of possible values of *q* if $3q = p+6$?

☐ A. $\{3,5,7\}$

☐ B. $\{4,5,6\}$

☐ C. $\{5,7,9\}$

☐ D. $\{6,8,10\}$

☐ E. $\{7,9,11\}$

31) If $y = 36$, then which of the following equations is correct?

☐ A. $y+20 = 60$

☐ B. $2y = 72$

☐ C. $4y = 140$

☐ D. $\frac{y}{3} = 15$

☐ E. $\frac{y}{4} = 10$

10.1 Practices

32) Alice scored a mean of 90 per test in her first 6 tests. In her 7th test, she scored 97. What was Alice's mean score for the 7 tests?

- ☐ A. 85
- ☐ B. 87
- ☐ C. 89
- ☐ D. 91
- ☐ E. 93

33) The volume of a cube is less than 125 m^3. Which of the following can be the cube's side?

- ☐ A. 4.5 m
- ☐ B. 5 m
- ☐ C. 6 m
- ☐ D. 6.5 m
- ☐ E. 7 m

34) What is the area of an isosceles right triangle that has one leg measuring 10 cm?

- ☐ A. 25 cm^2
- ☐ B. 50 cm^2
- ☐ C. 75 cm^2
- ☐ D. 100 cm^2
- ☐ E. 125 cm^2

35) If $0.00104 = \frac{208}{y}$, what is the value of y?

- ☐ A. 2,000
- ☐ B. 20,000
- ☐ C. 200,000
- ☐ D. 2,000,000
- ☐ E. 20,000,000

36) A bag is filled with numbered cards from 1 to 20 and one is picked out at random. What is the probability that the card picked is a prime number?

- ☐ A. $\frac{7}{20}$

- B. $\frac{2}{5}$
- C. $\frac{9}{20}$
- D. $\frac{10}{20}$
- E. $\frac{11}{20}$

37) In the xy-plane, the points $(10,8)$ and $(8,6)$ are on line B. Which of the following points could also be on line B?
- A. $(14,13)$
- B. $(8,7)$
- C. $(3,2)$
- D. $(1,0)$
- E. $(12,10)$

38) If $h(x) = 3x^4 - x^3 + 4x$ and $k(x) = 3$, what is the value of $h(k(x))$?
- A. 184
- B. 181
- C. 278
- D. 228
- E. 272

39) Two angles with $(3x-2)°$ and $119°$ degrees are supplementary. Find the value of x.
- A. 21
- B. 35
- C. 44
- D. 46
- E. 48

40) How many different three-digit numbers can be formed from the digits 5, 7, and 8, if the numbers must be even and no digit can be repeated?
- A. 1
- B. 2
- C. 3

10.1 Practices

☐ D. 4

☐ E. 5

41) A rectangular concrete structure is 40 feet long, 12 feet wide, and 3 feet deep. What is the volume of the concrete in cubic feet?

☐ A. $1,440\ ft^3$

☐ B. $1,520\ ft^3$

☐ C. $1,620\ ft^3$

☐ D. $1,800\ ft^3$

☐ E. $2,000\ ft^3$

42) Simplify the expression $\frac{2y}{x} - \frac{y}{3x}$.

☐ A. $\frac{2y+4}{3x}$

☐ B. $\frac{3y-6}{3x}$

☐ C. $\frac{5y}{3x}$

☐ D. $\frac{6y}{3x}$

☐ E. $\frac{8y}{3x}$

43) Evaluate the expression $200(3+0.01)^2 - 200$.

☐ A. 1206.02

☐ B. 1412.08

☐ C. 1030.04

☐ D. 1824.16

☐ E. 1612.02

44) If 720 kg of vegetables are packed in 120 boxes, how many kilograms of vegetables will each box contain?

☐ A. 4 kg

☐ B. 5 kg

☐ C. 6 kg

☐ D. 7 kg

☐ E. 8 kg

45) Each number in a sequence is 3 more than thrice the number that comes just before it. If 243 is a number in the sequence, what number comes just before it?

- A. 60
- B. 70
- C. 80
- D. 90
- E. 100

46) Evaluate the expression $[5 \times (-30) + 10] - (-6) + [3 \times 6] \div 2$.

- A. 163
- B. 147
- C. −153
- D. −157
- E. −125

47) Solve for x: $3 + \frac{4x}{x-6} = \frac{6}{6-x}$.

- A. $\frac{10}{7}$
- B. $\frac{12}{7}$
- C. $\frac{6}{7}$
- D. $\frac{8}{7}$
- E. $\frac{9}{7}$

48) The width of a rectangle is 8 *cm* and its length is 20 *cm*. What is the perimeter of this rectangle?

- A. 56 *cm*
- B. 64 *cm*
- C. 72 *cm*
- D. 80 *cm*
- E. 96 *cm*

49) What is the value of the following expression? $4\frac{1}{2} + 3\frac{2}{8} + 2\frac{1}{4} + 6\frac{3}{4}$

- A. $15\frac{11}{16}$
- B. $16\frac{1}{2}$

10.1 Practices

- ☐ C. $16\frac{3}{4}$
- ☐ D. $16\frac{5}{8}$
- ☐ E. $16\frac{7}{8}$

50) A certain insect has a mass of 120 milligrams. What is the insect's mass in grams?

- ☐ A. 0.12
- ☐ B. 0.012
- ☐ C. 1.2
- ☐ D. 12
- ☐ E. 120

10.2 Answer Keys

1) C. 32
2) C. $30
3) B. −8
4) C. 144°
5) A. 20
6) E. 89.6
7) A. $0.8 \leq x < 6$
8) A. $7.5\sqrt{3}$
9) E. Graph E
10) B. 18 cm
11) B. 4
12) E. 6.35
13) D. 47
14) C. $y < 1\frac{8}{15}$
15) B. 58.8
16) E. 30
17) D.
18) D. 30
19) A. $\frac{3}{7}$
20) C. $22.5\,m^2$
21) B. 18
22) C. $625
23) D. $\frac{4}{9}$
24) E. $8.72\,cm^2$
25) C. 6.67%

26) B. 150
27) A. 162
28) E. 5
29) D. 9.12
30) B. {4,5,6}
31) B. $2y = 72$
32) D. 91
33) A. $4.5\,m$
34) B. $50\,cm^2$
35) C. 200,000
36) B. $\frac{2}{5}$
37) E. (12, 10)
38) D. 228
39) A. 21
40) B. 2
41) A. $1,440\,ft^3$
42) C. $\frac{5y}{3x}$
43) E. 1612.02
44) C. $6\,kg$
45) C. 80
46) E. −125
47) B. $\frac{12}{7}$
48) A. $56\,cm$
49) C. $16\frac{3}{4}$
50) A. 0.12

10.3 Answers with Explanation

1) Let y be the capacity of the yellow container. Then the green container's capacity is $1.25y$:

$$1.25y = 40.$$

Solving for y gives:

$$y = \frac{40}{1.25} = 32.$$

Therefore, the yellow container can hold 32 balls, making option C correct.

2) Let the cost of the shoes be s. The cost of the dress is $2s - 15$:

$$2s - 15 = 45.$$

Solving for s gives $s = 30$. Therefore, the shoes cost \$30 and the option C is correct.

3) The value of $-\frac{45}{6}$ is -7.5. The greatest integer less than this value is -8, which is option B.

4) Let the angles be x, $2x$, $3x$, and $4x$. The sum of angles in a quadrilateral is $360°$:

$$x + 2x + 3x + 4x = 360° \Rightarrow 10x = 360.$$

Solving for x gives $x = 36°$, and the largest angle is $4x = 144°$, which makes option C correct.

5) The two angles shown in the figure are supplementary. So:

$$2x + 4x + 60 = 180 \Rightarrow 6x = 120 \Rightarrow x = 20°.$$

Therefore, the correct answer is option A.

6) The total sum of scores with the error is 60×90. Correcting the misread score:

$$\text{Corrected Sum} = (60 \times 90) - 98 + 74 = 5376.$$

The corrected mean is:
$$\text{Corrected Mean} = \frac{\text{Corrected Sum}}{60} = \frac{5376}{60} = 89.6.$$

Therefore, the corrected mean is 89.6, making option E correct.

7) To solve the compound inequality, adding 10 to all parts and then dividing by 5, gives:
$$-6 \leq 5x - 10 < 20 \Rightarrow 4 \leq 5x < 30 \Rightarrow 0.8 \leq x < 6.$$

Therefore, the option A is correct.

8) Since $a = 2b$, we have:
$$a + b = 3\sqrt{3} \Rightarrow 2b + b = 3\sqrt{3} \Rightarrow 3b = 3\sqrt{3} \Rightarrow b = \sqrt{3} \Rightarrow a = 2\sqrt{3}.$$

Area of the shaded trapezoid is:
$$\text{Area} = \frac{1}{2} \times (2\sqrt{3} + 3\sqrt{3}) \times 3 = 7.5\sqrt{3}.$$

Therefore, the correct answer is option A.

9) A linear equation describes a connection between variables x and y, typically expressed as $y = mx + b$. When considering a non-proportional linear relationship, it is represented by the equation $y = mx + b$, with the condition $b \neq 0$. This results in a linear graph that does not intersect the origin. Specifically, in graph E, the line depicted does not intersect the origin.

10) The triangles ABE and ACD are similar due to their equal angles. Therefore, the corresponding sides are proportional. We can set up the following proportion:
$$\frac{AE}{AD} = \frac{BE}{CD}.$$

Substituting the given lengths, we obtain:
$$\frac{x}{x+6} = \frac{9}{12} \Rightarrow 9x + 54 = 12x \Rightarrow 3x = 54 \Rightarrow x = 18.$$

10.3 Answers with Explanation

This calculation shows that x is 18 cm, option B.

11) Simple interest is calculated as Interest = Principal × Rate × Time. For $720 interest:

$$720 = 4500 \times 0.04 \times \text{Time}.$$

Solving for Time gives Time = 4 years, which makes option B correct.

12) Total kilograms picked:
$$2\frac{3}{5} + 3\frac{3}{4} = 2.6 + 3.75 = 6.35.$$

Therefore, Emily and John picked a total of 6.35 kilograms of oranges, which is option E.

13) The difference between $2n$ and $2m$ for even integers is the same as between n and m for all integers:

$$n - m = \frac{92}{2} = 46.$$

Including both ends of the list, the total count is:

$$46 + 1 = 47.$$

Therefore, the measure of list is 47, making option D correct.

14) To ensure that y is the shortest, it must be less than a third of the total length:

$$y < \frac{4\frac{3}{5}}{3} = \frac{23}{5} \times \frac{1}{3} = \frac{23}{15} = 1\frac{8}{15}.$$

Therefore, the shortest piece has a length less than $1\frac{8}{15}$, making option C correct.

15) Ben's total grade points are:
$$7 \times 84 = 588.$$

Jessica's mean grade is the total grade points divided by the number of courses:

$$\text{Mean Grade} = \frac{588}{10} = 58.8.$$

Therefore, Jessica's mean grade is 58.8 which is option B.

16) From the ratios, we have $a:b = 3:4$ and $b:c = 4:7$. Therefore, $a = \frac{3}{4}b$ and $c = \frac{7}{4}b$. The sum is:

$$\frac{3}{4}b + b + \frac{7}{4}b = 105.$$

Solving for b gives:

$$\frac{14}{4}b = 105 \Rightarrow b = 105 \times \frac{4}{14} = 30.$$

Therefore, the value of b is 30, making option E correct.

17) To solve the inequality $-4 < -2x \leq 6$, we divide the entire inequality by -2 and reverse the inequality signs (since we are dividing by a negative number):

$$\frac{-4}{-2} > x \geq \frac{6}{-2}.$$

Simplifying each part of the inequality, we get:

$$2 > x \geq -3.$$

So, the solution to the inequality is $-3 \leq x < 2$, which is option D.

18) The set T includes the odd integers from 11 to 49. These numbers form an arithmetic sequence with a common difference of 2. The mean is the average of the first and last terms:

$$\text{Mean} = \frac{11 + 49}{2} = 30.$$

Therefore, the mean of the members in T is 30, making option D correct.

19) Squaring both sides of the equation gives:

$$3z = 7w.$$

10.3 Answers with Explanation

Solving for the ratio $\frac{w}{z}$ yields:
$$\frac{w}{z} = \frac{3}{7}.$$

Therefore, the ration of w to z is $3:7$ or $\frac{3}{7}$, making option A correct.

20) First, calculate the area of triangle MBN:

$$\text{Area of } \triangle MBN = \frac{1}{2} \times \text{Base} \times \text{Height} = \frac{1}{2} \times 3 \times 3 = 4.5 \text{ m}^2.$$

Next, calculate the area of triangle NCD:

$$\text{Area of } \triangle NCD = \frac{1}{2} \times \text{Base} \times \text{Height} = \frac{1}{2} \times 3 \times 6 = 9 \text{ m}^2.$$

The area of the square ABCD is:

$$\text{Area of square} = (\text{side})^2 = 6 \times 6 = 36 \text{ m}^2.$$

Subtract the areas of the triangles from the area of the square to find the area of AMND:

$$\text{Area of AMND} = \text{Area of square} - (\text{Area of } \triangle MBN + \text{Area of } \triangle NCD)$$

$$= 36 - (4.5 + 9) = 36 - 13.5 = 22.5 \text{ m}^2.$$

Therefore, the area of the polygon AMND is 22.5 m^2, making option C correct.

21) If 70% cannot swim, then 30% can swim. The number of students who can swim is 30% of 60:

$$\text{Number of swimming students} = \frac{30}{100} \times 60 = 18.$$

Therefore, option B is correct.

22) Each person gets an equal share of the total amount:

$$\text{Amount per person} = \frac{\$10,000}{16} = \$625.$$

Therefore, C is the correct option.

23) The total number of sticks is $8+6+4=18$. The probability of picking a green stick is:

$$\text{Probability of green stick} = \frac{8}{18} = \frac{4}{9}.$$

Which is option D.

24) First, calculate the area of the circle:

$$\text{Area of the circle} = \pi r^2 = \pi \times 4^2 = 16\pi \text{ cm}^2.$$

Next, calculate the area of the regular hexagon. Each side of hexagon is equal to the radius of the circle, which is 4 cm. Each of the six triangles in the hexagon is equilateral with side 4 cm. The area of one such triangle is:

$$\text{Area of one triangle} = \frac{\sqrt{3}}{4} \times (\text{side})^2 = \frac{\sqrt{3}}{4} \times 4^2 = 4\sqrt{3} \text{ cm}^2.$$

Since the hexagon consists of 6 such triangles, the total area of the hexagon is:

$$\text{Area of hexagon} = 6 \times 4\sqrt{3} = 24\sqrt{3} \text{ cm}^2.$$

Now, subtract the area of the hexagon from the area of the circle to find the shaded area:

$$\text{Shaded area} = \text{Area of the circle} - \text{Area of hexagon} = 16\pi - 24\sqrt{3} \text{ cm}^2.$$

Using the approximations $\pi \approx 3.14$ and $\sqrt{3} \approx 1.73$, we get:

$$\text{Shaded area} \approx 16 \times 3.14 - 24 \times 1.73 = 50.24 - 41.52 = 8.72 \text{ cm}^2.$$

Therefore, the correct answer is E.

25) The percent increase is calculated by the formula:

$$\text{Percent Increase} = \frac{\text{New Price} - \text{Old Price}}{\text{Old Price}} \times 100\%.$$

10.3 Answers with Explanation

Thus, Percent Increase $= \frac{8.00-7.50}{7.50} \times 100\% \approx 6.67\%$, which is option C.

26) Let x denote the red marbles. Using the ratio, we have:
$$\frac{5}{6} = \frac{x}{180}.$$

Solving for x, we get:
$$x = \frac{180 \times 5}{6} = 150.$$

So, the correct option is B. 150.

27) Let the number be x. Then:
$$\frac{7}{9}x = 126.$$

Solving for x:
$$x = \frac{126 \times 9}{7} = 162.$$

Therefore, the number is 162, making option A correct.

28) The ratio of banana puree to strawberry puree is:
$$\frac{\frac{2}{5}}{\frac{2}{25}} = \frac{2}{5} \times \frac{25}{2} = 5.$$

Therefore, the mixture contains 5 cup of banana puree per cup of strawberry puree, making option E correct.

29) The diameter of the circle is equal to the diagonal of the square, so the radius r is:
$$r = \frac{\sqrt{4^2+4^2}}{2} = \frac{\sqrt{32}}{2} = \frac{4\sqrt{2}}{2} = 2\sqrt{2}.$$

The area of the circle is $A_{circle} = \pi r^2$ and the area of the square is $A_{square} = (side)^2$. The shaded area A_{shaded} is:

$$A_{shaded} = A_{circle} - A_{square} = 3.14 \times (2\sqrt{2})^2 - 4^2 = 3.14 \times 8 - 16 = 9.12.$$

Therefore, the correct answer is option D.

30) Substitute each value of p into the equation $3q = p+6$ and solve for q:

$$3q = 6+6,\ 3q = 9+6,\ 3q = 12+6.$$

Solving each gives the corresponding values for q: $4, 5, 6$, making option B correct.

31) To find the correct equation, substitute $y = 36$ into each option and evaluate:

A. $y + 20 = 60 \Rightarrow 36 + 20 = 56$, which is not equal to 60.

B. $2y = 72 \Rightarrow 2 \times 36 = 72$, which is correct.

C. $4y = 140 \Rightarrow 4 \times 36 = 144$, which is not equal to 140.

D. $\frac{y}{3} = 15 \Rightarrow \frac{36}{3} = 12$, which is not equal to 15.

E. $\frac{y}{4} = 10 \Rightarrow \frac{36}{4} = 9$, which is not equal to 10.

Therefore, the only true statement after substituting $y = 36$ is option B, $2y = 72$.

32) Alice's total score for the first 6 tests is 6×90. Adding the 7th test score:

$$\text{Total score} = 6 \times 90 + 97 = 637.$$

The mean score for 7 tests is:

$$\text{Mean score} = \frac{\text{Total score}}{7} = \frac{637}{7} = 91.$$

Therefore, the mean score for the 7 tests is 91, making option D correct.

33) For the cube's volume to be less than $125\ m^3$, its side must be less than $\sqrt[3]{125} = 5$. Therefore, the correct option is A.

34) Let a be the length of the legs of the right triangle:

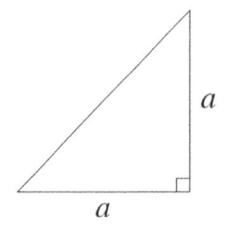

10.3 Answers with Explanation

The area A of an isosceles right triangle is given by $A = \frac{1}{2} \times a^2$:

$$A = \frac{1}{2} \times 10^2 = 50 \ cm^2.$$

Therefore, the area is 50 cm^2, making option B correct.

35) Solving the equation for y:

$$y = \frac{208}{0.00104} = 200,000.$$

Therefore, the value of y is 200,000, which makes option C correct.

36) First, identify the prime numbers from 1 to 20: $2, 3, 5, 7, 11, 13, 17, 19$. There are a total of 8 prime numbers.

Since there are 20 cards in total, the probability of picking a prime number is:

$$\text{Probability} = \frac{\text{Number of prime cards}}{\text{Total number of cards}} = \frac{8}{20} = \frac{2}{5}.$$

Therefore, the correct answer is option B, $\frac{8}{20}$ or $\frac{2}{5}$.

37) To find the equation of line B, we start by calculating its slope. Using the points $(10, 8)$ and $(8, 6)$ we have:

$$\text{Slope} = \frac{y_2 - y_1}{x_2 - x_1} = \frac{8 - 6}{10 - 8} = \frac{2}{2} = 1.$$

With the slope known, we use the point-slope form of the equation of a line, $y - y_1 = m(x - x_1)$, where m is the slope and (x_1, y_1) is a point on the line. Using the point $(8, 6)$ and the slope 1:

$$y - 6 = 1(x - 8) \Rightarrow y = x - 2.$$

This equation represents line B. To determine which of the given points lies on this line, substitute the x and y values of each point into the equation and see if it holds true.

- For A, $(14, 13)$, $13 \neq 14 - 2$.
- For B, $(8, 7)$, $7 \neq 8 - 2$.
- For C, $(3, 2)$, $2 \neq 3 - 2$.
- For D, $(1, 0)$, $0 \neq 1 - 2$.

- For E, $(12, 10)$, $10 = 12 - 2$ (True).

Therefore, the only point that satisfies the equation of line B is option E.

38) Substitute $k(x) = 3$ into $h(x)$:

$$h(k(x)) = h(3) = 3 \times 3^4 - 3^3 + 4 \times 3 = 228.$$

Therefore, the correct option is D.

39) Supplementary angles add up to $180°$:

$$(3x - 2) + 119 = 180 \Rightarrow 3x = 180 - 119 + 2 \Rightarrow 3x = 63 \Rightarrow x = 21.$$

Therefore, The value of x is 21, making option A correct.

40) To form a three-digit even number using the digits $5, 7$, and 8, without repeating any digit, certain conditions must be met:

1. The unit's place must be occupied by an even digit to ensure the number is even. In this case, the only even digit available is 8.

2. The hundred's place can be occupied by either 5 or 7. This leaves two choices for the hundred's place.

3. The ten's place will then be occupied by the remaining digit (which was not used in the hundred's place).

With these constraints, the only possible combinations are 578 and 758. Here's how these numbers are formed: Thus, there are only 2 different three-digit even numbers that can be formed under these conditions, making option B correct.

41) The volume V of the structure is calculated by the formula:

$$V = \text{length} \times \text{width} \times \text{depth}.$$

Thus:

$$V = 40 \ ft \times 12 \ ft \times 3 \ ft = 1,440 \ ft^3.$$

Therefore, the correct answer is option A.

10.3 Answers with Explanation 257

42) Combine the fractions:
$$\frac{2y}{x} - \frac{y}{3x} = \frac{6y}{3x} - \frac{y}{3x} = \frac{6y-y}{3x} = \frac{5y}{3x}.$$

Therefore, the correct answer is option C.

43) Expand and simplify the expression:

$$200(3+0.01)^2 - 200 = 200(3.01)^2 - 200$$
$$= 200(9.0601) - 200$$
$$= 1812.02 - 200$$
$$= 1612.02.$$

Therefore, the correct answer is option E.

44) Divide the total weight by the number of boxes:

$$\text{Vegetables per box} = \frac{720\ kg}{120} = 6\ kg.$$

Therefore, each box contains 6 kilograms vegetables, making option C correct.

45) Let the preceding number be y. The formula for the sequence is $243 = 3y + 3$. Solving for y:

$$3y = 243 - 3 \Rightarrow y = \frac{240}{3} \Rightarrow y = 80.$$

Therefore, the correct answer is option C.

46) Compute each part of the expression step by step:

$$[5 \times (-30) + 10] - (-6) + [3 \times 6] \div 2 = [-150 + 10] + 6 + 9$$
$$= -140 + 6 + 9$$
$$= -134 + 9$$
$$= -125.$$

Thus, the correct option is E.

47) Multiply both sides by $(x-6)$ to clear the denominators and solve for x:

$$3(x-6)+4x = 6(-1) \Rightarrow 3x-18+4x = -6 \Rightarrow 7x = 12 \Rightarrow x = \frac{12}{7}.$$

Thus, the correct option is B.

48) The perimeter P of a rectangle is given by $P = 2 \times (\text{width} + \text{length})$:

$$P = 2 \times (8 \text{ } cm + 20 \text{ } cm) = 56 \text{ } cm.$$

Therefore, the perimeter of rectangle is 56 cm, making option A correct.

49) Add the mixed numbers:

$$4\frac{1}{2}+3\frac{2}{8}+2\frac{1}{4}+6\frac{3}{4} = (4+3+2+6)+(\frac{1}{2}+\frac{2}{8}+\frac{1}{4}+\frac{3}{4}) = 15+1\frac{3}{4} = 16\frac{3}{4}.$$

Therefore, the correct answer is option C.

50) Convert milligrams to grams by dividing by 1000 (since 1000 milligrams make a gram):

$$120 \text{ mg} = \frac{120}{1000} \text{ g} = 0.12 \text{ g}.$$

Therefore, the correct option is A.

11. Practice Test 10

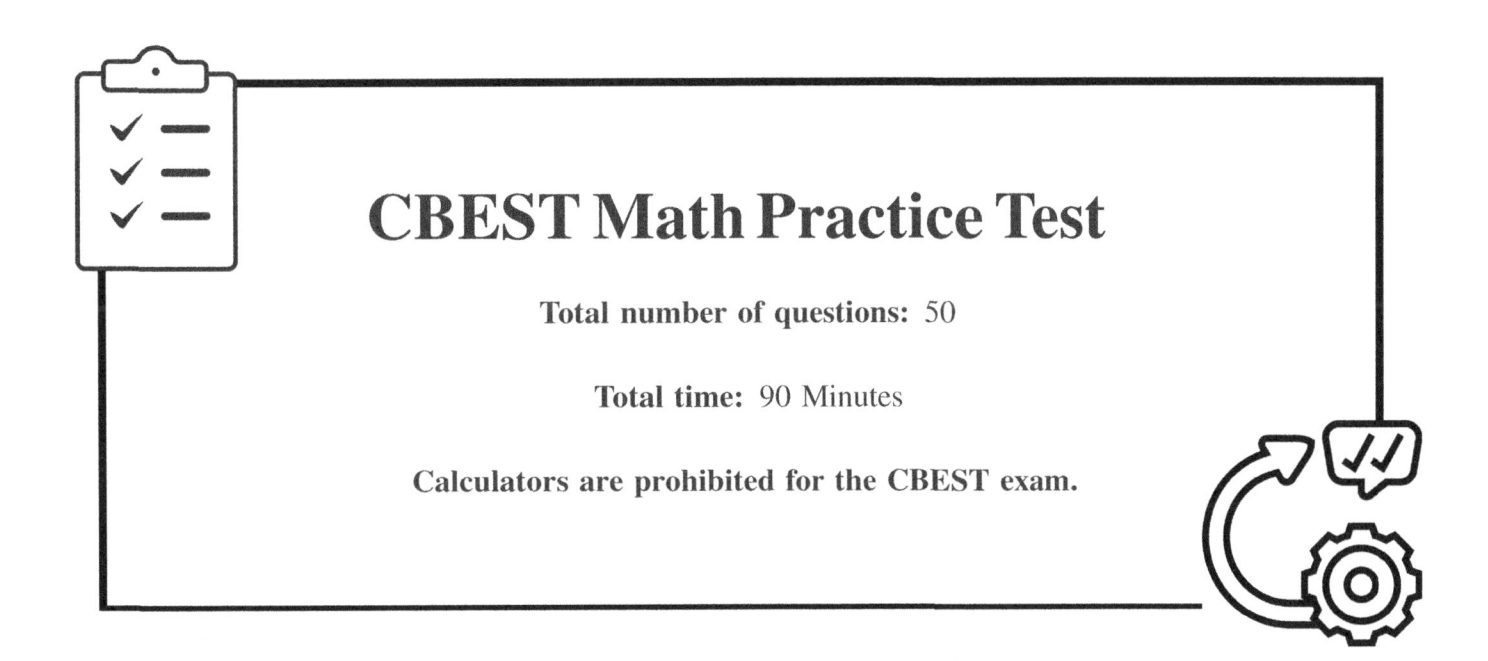

CBEST Math Practice Test

Total number of questions: 50

Total time: 90 Minutes

Calculators are prohibited for the CBEST exam.

11.1 Practices

1) A computer was originally priced at $800.00 and was on sale for 20% off. Lisa, an employee, received a 15% discount on top of the sale price. How much did Lisa pay for the computer?

☐ A. $544.00

☐ B. $576.00

☐ C. $680.00

☐ D. $700.00

2) In a group of 30 students, there are 20 boys and the rest are girls. What is the ratio of the number of girls to the number of boys?

- ☐ A. 1 : 2
- ☐ B. 1 : 3
- ☐ C. 1 : 4
- ☐ D. 2 : 5
- ☐ E. 1 : 5

3) The shaded sector of the circle shown below has an area of 75π square meters. What is the circumference of the circle?

- ☐ A. 40π meters
- ☐ B. 60π meters
- ☐ C. 120π meters
- ☐ D. 150π meters
- ☐ E. 180π meters

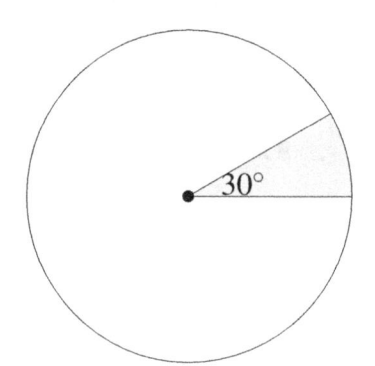

4) Which of the following is a factor of 72?

- ☐ A. 7
- ☐ B. 9
- ☐ C. 10
- ☐ D. 14
- ☐ E. 16

5) By what percent did the cost of a laptop increase if its price was increased from $520.00 to $598.00?

- ☐ A. 12%
- ☐ B. 14%
- ☐ C. 15%
- ☐ D. 18%
- ☐ E. 20%

11.1 Practices

6) The greatest common factor of 48 and x is 12. How many possible values for x are greater than 15 and less than 70?

- ☐ A. 5
- ☐ B. 4
- ☐ C. 3
- ☐ D. 2
- ☐ E. 1

7) A bag contains 8 apple candies, 5 lemon candies, and 4 grape candies. If Jenna selects 2 candies at random from this bag, without replacement, what is the probability that none of the candies are grape candies?

- ☐ A. $\frac{1}{17}$
- ☐ B. $\frac{3}{17}$
- ☐ C. $\frac{5}{17}$
- ☐ D. $\frac{17}{68}$
- ☐ E. $\frac{39}{68}$

8) How many integers are between $\frac{9}{3}$ and $\frac{33}{4}$?

- ☐ A. 5
- ☐ B. 7
- ☐ C. 8
- ☐ D. 9
- ☐ E. 10

9) In a certain region, the sales tax rate increased from 6% to 6.5%. What was the increase in the sales tax on a $300 purchase?

- ☐ A. $1.00
- ☐ B. $1.50
- ☐ C. $1.80
- ☐ D. $2.00
- ☐ E. $2.50

10) Triangle PQR is graphed on a coordinate grid with vertices at P (2, 3), Q (4, −5), and R (−6, −8). Triangle PQR is reflected over the y-axis to create triangle $P'Q'R'$. Which pair represents the coordinate of R'?

- A. (6, −8)
- B. (6, 8)
- C. (−6, −8)
- D. (−6, 8)
- E. (8, −6)

11) Which of the following is the solution of the inequality?

$$3x - 6 < 15x + 4.5 - 6.75x.$$

- A. $x < -2$
- B. $x \leq 2$
- C. $x > -2$
- D. $x \leq 3$
- E. $x \geq -3$

12) What is the volume of the following triangular prism?

- A. $20\ m^3$
- B. $18\ m^3$
- C. $16\ m^3$
- D. $14\ m^3$
- E. $12\ m^3$

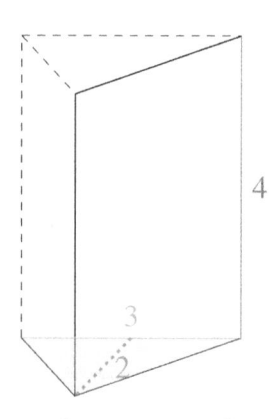

13) Make a list of all possible products of 2 different numbers in {2, 3, 8, 7, 9}. What fraction of the products are odd?

- A. $\frac{1}{5}$
- B. $\frac{2}{9}$
- C. $\frac{4}{10}$
- D. $\frac{3}{10}$
- E. $\frac{5}{15}$

14) Which of the following equations has a graph that is a straight line?

11.1 Practices

- [ ] A. $y = 2x^2 + 4$
- [ ] B. $x^2 + y = 5$
- [ ] C. $3x - y = 7$
- [ ] D. $5x + xy = 3$
- [ ] E. $y + 3 = x^2$

15) How many odd numbers are in the range from $4n$ up to and including $4n + 6$?

- [ ] A. 1
- [ ] B. 2
- [ ] C. 3
- [ ] D. 4
- [ ] E. 5

16) What is the value of y in the following equation?

$$\frac{3}{4}y - \frac{1}{8} = \frac{1}{2}.$$

- [ ] A. 1
- [ ] B. $\frac{2}{3}$
- [ ] C. $\frac{3}{4}$
- [ ] D. $\frac{5}{6}$
- [ ] E. $\frac{7}{8}$

17) A credit union offers 3% simple interest on a fixed deposit. If you deposit $20,000, how much interest will you earn in five years?

- [ ] A. $1,000
- [ ] B. $3,000
- [ ] C. $2,000
- [ ] D. $6,000
- [ ] E. $10,000

18) If $3x - 7y = 21$, what is x in terms of y?

- [ ] A. $x = \frac{7}{3}y + 7$

- B. $x = \frac{3}{7}y + 21$
- C. $x = -\frac{7}{3}y - 7$
- D. $x = -\frac{7}{3}y + 7$
- E. $x = \frac{3}{7}y - 21$

19) For what value of y is the proportion true? $y : 50 = 25 : 40$

- A. 20
- B. 31.25
- C. 35
- D. 40
- E. 45

20) In a survey, 80% of the participants chose option B. If 40 participants chose an option other than B, what was the total number of participants?

- A. 100
- B. 160
- C. 200
- D. 250
- E. 320

21) Which percentage is closest in value to 0.0123?

- A. 1.2%
- B. 1.23%
- C. 1.5%
- D. 2%
- E. 2.3%

22) Calculate the surface area of a cone with a radius of 5 inches and a slant height of 10 inches.

11.1 Practices

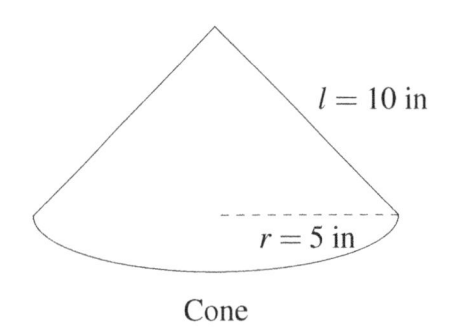

☐ A. $75\pi\ in^2$

☐ B. $100\pi\ in^2$

☐ C. $150\pi\ in^2$

☐ D. $200\pi\ in^2$

☐ E. $250\pi\ in^2$

$l = 10$ in

$r = 5$ in

Cone

23) A bus travels 800 miles from Los Angeles to Denver. It covers the first 160 miles in 3 hours. If the bus maintains this rate, how many more hours will it take to reach Denver? Round your answer to the nearest whole hour.

☐ A. 10

☐ B. 15

☐ C. 14

☐ D. 16

☐ E. 12

24) Which of the following graphs represents the solution of the inequality $|2x - 4| < 6$?

☐ A.

☐ B.

☐ C.

☐ D.

☐ E.

25) In the xy-plane, the line determined by the points $(8, n)$ and $(n, 16)$ passes through the origin. Which of the following could be the value of n?

☐ A. $8\sqrt{2}$

☐ B. $10\sqrt{2}$

☐ C. $10\sqrt{3}$

☐ D. $\sqrt{3}$

☐ E. $\sqrt{7}$

26) What is the average of the circumference of a circle with diameter 14 and the circumference of a square with side length 10? (Use $\pi = 3$)

☐ A. 38

☐ B. 39

☐ C. 40

☐ D. 41

☐ E. 42

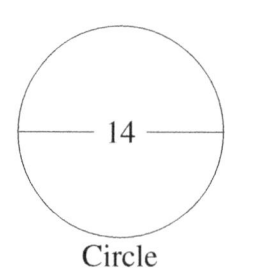

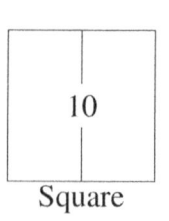

Circle (14)   Square (10)

27) The sale price of a television is $2,475, which is 25% off the original price. What is the original price of the television?

☐ A. $3,000

☐ B. $3,300

☐ C. $3,625

☐ D. $2,475

☐ E. $2,030

28) The radius of a semi-circular arch is 20 cm. What is the perimeter of the shape? ($\pi = 3.14$)

☐ A. 82.8 *cm*

☐ B. 54.28 *cm*

☐ C. 140.28 *cm*

☐ D. 102.8 *cm*

☐ E. 62.8 *cm*

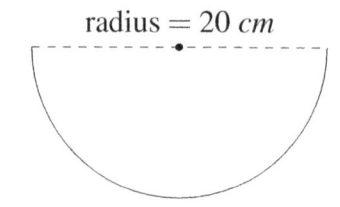

radius = 20 *cm*

29) In a scale diagram, 0.2 inch represents 100 feet. How many inches represent 5 feet?

☐ A. 0.001 *in*

☐ B. 0.005 *in*

☐ C. 0.01 *in*

☐ D. 0.02 *in*

☐ E. 0.1 *in*

30) If $\frac{4}{9}$ of X is 72, what is $\frac{3}{4}$ of X?

☐ A. 145.6

☐ B. 154.5

☐ C. 163.5

☐ D. 130.2

☐ E. 121.5

31) A motorcycle travels at a speed of 60 miles per hour. How far will it travel in 6 hours?

☐ A. 360

☐ B. 330

☐ C. 300

☐ D. 270

☐ E. 240

32) If Lucy spent $40 on books, and she spent 20% of the price for a bag, how much did she spend in total?

☐ A. $44

☐ B. $46

☐ C. $48

☐ D. $50

☐ E. $52

33) Which of the following numbers does not have any factors that include the smallest factor (other than 1) of 77?

☐ A. 21

☐ B. 35

☐ C. 42

☐ D. 49

☐ E. 54

34) Evaluate the expression: $\frac{5^2+4^2+(-6)^2}{(12+15-18)^2}$

☐ A. $\frac{77}{81}$

☐ B. $-\frac{77}{81}$

☐ C. 1

☐ D. 77

☐ E. 81

35) Angles C and D are supplementary. The measure of angle C is 3 times the measure of angle D. What is the measure of angle C in degrees?

- A. 150°
- B. 135°
- C. 165°
- D. 175°
- E. 180°

36) If $z = -3$ in the following equation, what is the value of w? $3z + 4 = \frac{w+8}{4}$

- A. -12
- B. -14
- C. -16
- D. -18
- E. -28

37) Jacob is 6 feet 10.5 inches tall, and Emma is 5 feet 4 inches tall. What is the difference in height, in inches, between Emma and Jacob?

- A. 18.5
- B. 19.5
- C. 20.5
- D. 21.5
- E. 22.5

38) Simplify: $\dfrac{\left(\frac{50(y+2)}{5} - 15\right)}{15}$

- A. $\frac{2}{5}y$
- B. $\frac{3}{5}y + \frac{2}{3}$
- C. $\frac{2}{3}y + \frac{1}{3}$
- D. $\frac{4}{7}y + 1$
- E. $\frac{1}{3}y + 2$

39) Yesterday, Max wrote 15% of his essay. Today, he wrote another 20% of the entire essay. What fraction of the essay is left for him to write?

- A. $\frac{13}{20}$
- B. $\frac{14}{20}$

11.1 Practices

- [ ] C. $\frac{15}{20}$
- [ ] D. $\frac{17}{25}$
- [ ] E. $\frac{18}{25}$

40) In a box of red and green markers, the ratio of red markers to green markers is 3 : 4. If the box contains 12 green markers, how many red markers are there?

- [ ] A. 6
- [ ] B. 9
- [ ] C. 12
- [ ] D. 15
- [ ] E. 18

41) What decimal is equivalent to $-\frac{8}{12}$?

- [ ] A. $-0.\overline{6}$
- [ ] B. -0.67
- [ ] C. -0.68
- [ ] D. $-0.5\overline{6}$
- [ ] E. $-0.7\overline{1}$

42) The area of a circle is 121π. What is the diameter of the circle?

- [ ] A. 11
- [ ] B. 12
- [ ] C. 10
- [ ] D. 22
- [ ] E. 24

43) Ten years ago, John was four times as old as Sarah was. If Sarah is 15 years old now, how old is John?

- [ ] A. 20
- [ ] B. 35
- [ ] C. 30
- [ ] D. 45
- [ ] E. 50

44) How many positive odd factors of 120 are greater than 20 and less than 100?

☐ A. 0

☐ B. 1

☐ C. 2

☐ D. 3

☐ E. 4

45) The ratio of two sides of a rectangle is 3:4. If its perimeter is 56 cm, find the length of its sides.

☐ A. 9 *cm* and 12 *cm*

☐ B. 12 *cm* and 16 *cm*

☐ C. 14 *cm* and 18 *cm*

☐ D. 15 *cm* and 20 *cm*

☐ E. 18 *cm* and 24 *cm*

46) What is the value of *y* in the following equation? $\frac{5}{6}(y-3) = 4(\frac{1}{8}y - 1)$

☐ A. 0.5

☐ B. −0.5

☐ C. −4.5

☐ D. 6.2

☐ E. 4.5

47) If *z* can be any integer, what is the greatest possible value of the expression $3 - z^2$?

☐ A. −1

☐ B. 3

☐ C. 1

☐ D. 2

☐ E. 4

48) A bag contains 5 red, 7 blue, 9 white, and 12 yellow balls. If one ball is picked at random, what is the probability that it will be red?

☐ A. $\frac{1}{6}$

☐ B. $\frac{5}{11}$

- ☐ C. $\frac{5}{33}$
- ☐ D. $\frac{7}{24}$
- ☐ E. $\frac{11}{33}$

49) In the given figure below, F is the midpoint of EH and G is the midpoint of FH. Which segment has length $2y - 3x$ centimeters?

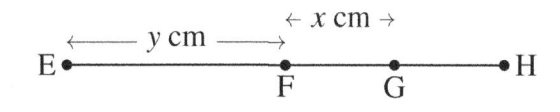

- ☐ A. GH
- ☐ B. EF
- ☐ C. FH
- ☐ D. EG
- ☐ E. EH

50) Jack answered 15 out of 60 questions on a test incorrectly. What percentage of the questions did he answer correctly?

- ☐ A. 25%
- ☐ B. 50%
- ☐ C. 75%
- ☐ D. 85%
- ☐ E. 90%

11.2 Answer Keys

1) A. $544.00
2) A. 1 : 2
3) B. 60π meters
4) B. 9
5) C. 15%
6) D. 2
7) E. $\frac{39}{68}$
8) A. 5
9) B. $1.50
10) A. $(6, -8)$
11) C. $x > -2$
12) E. 12 m^3
13) D. $\frac{3}{10}$
14) C. $3x - y = 7$
15) C. 3
16) D. $\frac{5}{6}$
17) B. $3,000
18) A. $x = \frac{7}{3}y + 7$
19) B. 31.25
20) C. 200
21) B. 1.23%
22) A. 75π in^2
23) E. 12
24) A. $-1 < x < 5$
25) A. $8\sqrt{2}$

26) D. 41
27) B. $3,300
28) D. 102.8 cm
29) C. 0.01 in
30) E. 121.5
31) A. 360
32) C. $48
33) E. 54
34) A. $\frac{77}{81}$
35) B. $135°$
36) E. -28
37) A. 18.5
38) C. $\frac{2}{3}y + \frac{1}{3}$
39) A. $\frac{13}{20}$
40) B. 9
41) A. $-0.\overline{6}$
42) D. 22
43) C. 30
44) A. 0
45) B. 12 cm and 16 cm
46) C. -4.5
47) B. 3
48) C. $\frac{5}{33}$
49) A. GH
50) C. 75%

11.3 Answers with Explanation

1) First, calculate the sale price after the initial 20% discount:

$$800 \times (1 - 0.20) = 800 \times 0.80 = \$640.$$

Then, apply the 15% employee discount to the sale price:

$$640 \times (1 - 0.15) = 640 \times 0.85 = \$544.00.$$

Therefore, Lisa paid $544.00 for the computer, making option A correct.

2) The number of girls is $30 - 20 = 10$. The ratio of girls to boys is:

$$\frac{10}{20} = \frac{1}{2}.$$

Therefore, the ratio of girls to boys is 1 : 2, making option A correct.

3) The area of shaded sector is one twelfth of the area of the entier circle. Thus, the total area of the circle is $12 \times 75\pi = 900\pi$ square meters. The radius can then be calculated using the area formula $A = \pi r^2$:

$$900\pi = \pi r^2 \Rightarrow r^2 = 900 \Rightarrow r = 30 \text{ meters}.$$

The circumference is given by $C = 2\pi r = 2\pi \times 30 = 60\pi$ meters, making option B correct.

4) To find a factor of 72, look for a number that divides 72 without leaving a remainder. Among the options, 9 divides 72 as $72 \div 9 = 8$. Thus, 9 is a factor of 72, making option B correct.

5) First calculate the absolute increase:

$$598 - 520 = 78.$$

Then calculate the percentage increase:

$$\left(\frac{78}{520}\right) \times 100\% = 15\%.$$

Thus, the price of the laptop increased by 15%, making option C correct.

6) To find the possible values of x, list the multiples of 12 that are greater than 15 and less than 70:

$$24, 36, 48, 60.$$

Among thsese numbers, 36 and 60 are the only values that meet the criteria and also have 12 as their greatest common factor with 48. Thus, there are 2 possible values for x, making option D correct.

7) The total number of candies is $8+5+4=17$. The probability that the first candy is not grape is $\frac{13}{17}$ (apple or lemon), and the probability that the second candy is not grape, after one non-grape candy has been removed, is $\frac{12}{16}$. Thus, the probability that none of the candies are grape candies is:

$$\frac{13}{17} \times \frac{12}{16} = \frac{39}{68}.$$

Therefore, the probability is $\frac{39}{68}$, making option E correct.

8) First, convert the fractions to integers. $\frac{9}{3}=3$ and $\frac{33}{4}$ is just over 8. Therefore, The integers between 3 and $\frac{33}{4}$ (exclusive) are:

$$4, 5, 6, 7, 8,$$

making option A correct.

9) The increase in tax rate is $6.5\% - 6\% = 0.5\%$. The increase in sales tax on a $300 purchase is:

$$300 \times 0.005 = \$1.50.$$

Therefore, the increase in sales tax is $1.50, making option B correct.

10) Reflecting a point over the y-axis changes the sign of the x-coordinate but keeps the y-coordinate the same. Thus, the coordinates of R' are $(6, -8)$, making option A correct.

11) Simplify the inequality:

$$3x - 6 < 15x - 6.75x + 4.5 \Rightarrow 3x - 6 < 8.25x + 4.5.$$

11.3 Answers with Explanation

Then solve for x:

$$-6 - 4.5 < 8.25x - 3x \Rightarrow -10.5 < 5.25x \Rightarrow x > \frac{-10.5}{5.25} = -2.$$

Thus, the solution is $x > -2$, making option C correct.

12) The volume of a triangular prism is calculated as the base area times the height. For this prism:

$$\text{Volume} = \frac{1}{2} \times 2\,m \times 3\,m \times 4\,m = 12\,m^3.$$

Therefore, the volume of the prism is $12\,m^3$, making option E correct.

13) To find the fraction of products that are odd, we need to consider the properties of odd and even numbers in multiplication. The product of two numbers is odd only if both numbers are odd.

In the given set $\{2, 3, 8, 7, 9\}$, there are three odd numbers (3, 7, and 9) and two even numbers (2 and 8).

First, calculate the total number of distinct products that can be formed using two different numbers from the set. Since there are 5 numbers in the set, the total number of products is given by the combination formula for selecting 2 items from 5:

$$\text{Total number of products} = \binom{5}{2} = \frac{5 \times 4}{2 \times 1} = 10.$$

Next, determine the number of products that are odd. As stated earlier, a product is odd only if both factors are odd. There are 3 odd numbers in the set, so the number of odd products is the number of ways to choose 2 odd numbers from 3:

$$\text{Number of odd products} = \binom{3}{2} = \frac{3 \times 2}{2 \times 1} = 3.$$

Finally, calculate the fraction of products that are odd:

$$\text{Fraction of odd products} = \frac{\text{Number of odd products}}{\text{Total number of products}} = \frac{3}{10}.$$

Therefore, the fraction of products that are odd is $\frac{3}{10}$, which is option D.

14) Only linear equations in the form $y = mx + b$ or rearrangements of this form represent straight lines. Among the options, $3x - y = 7$ can be rearranged to the linear form, making option C correct.

15) Since $4n$ is even, the odd numbers in the range are $4n+1$, $4n+3$, and $4n+5$. Therefore, there are 3 odd numbers in the range, making option C correct.

16) Solving the equation:

$$\frac{3}{4}y = \frac{1}{2} + \frac{1}{8} = \frac{4+1}{8} = \frac{5}{8} \Rightarrow y = \frac{\frac{5}{8}}{\frac{3}{4}} = \frac{5}{8} \times \frac{4}{3} = \frac{5}{6}.$$

Thus, the value of y is $\frac{5}{6}$, making option D correct.

17) The interest earned is calculated as:

$$\text{Principal} \times \text{Rate} \times \text{Time} = 20000 \times 0.03 \times 5 = \$3000.$$

Thus, the interest earned over five years is $3000, making option B correct.

18) Rearranging the equation:

$$3x = 7y + 21 \Rightarrow x = \frac{7y+21}{3} = \frac{7}{3}y + 7.$$

Therefore, x in terms of y is $\frac{7}{3}y + 7$, making option A correct.

19) Set up the proportion:

$$\frac{y}{50} = \frac{25}{40}.$$

Solving for y:

$$y = \frac{25 \times 50}{40} = 31.25,$$

hence, the correct value for y is 31.25, making option B correct.

20) If 80% chose B, then 20% did not choose B. Let T be the total number of participants. Thus,

$$0.20 \times T = 40.$$

Solving for T,

$$T = \frac{40}{0.20} = 200.$$

11.3 Answers with Explanation

Therefore, the total number of participants was 200, making option C correct.

21) To convert a decimal to a percentage, multiply by 100%:

$$0.0123 \times 100\% = 1.23\%.$$

Thus, the closest percentage value to 0.0123 is 1.23%, making option B correct.

22) The surface area of a cone is given by the formula $\pi r(r+l)$, where r is the radius and l is the slant height. Therefore:

$$\text{Surface Area} = \pi \times 5 \times (5+10) = 75\pi \ in^2,$$

hence, the correct answer is option A, $75\pi \ in^2$.

23) First, find the rate of travel:

$$\text{Rate} = \frac{160 \text{ miles}}{3 \text{ hours}} \approx 53.33 \text{ miles/hour}.$$

Then, calculate the remaining time to travel $800 - 160 = 640$ miles:

$$\text{Time} = \frac{640 \text{ miles}}{53.33 \text{ miles/hour}} \approx 12 \text{ hours}.$$

Therefore, it will take approximately 12 more hours to reach Denver, making option E correct.

24) The solution to the inequality $|2x - 4| < 6$ can be found by considering the compound inequalities: $-6 < 2x - 4 < 6$. Solving these:

$$-6 < 2x - 4 < 6 \Rightarrow -2 < 2x < 10 \Rightarrow -1 < x < 5.$$

Therefore, the correct answer is option A.

25) The slope of the line through the points $(8, n)$ and $(n, 16)$ is:

$$\text{Slope} = \frac{16 - n}{n - 8}.$$

Thus, the equation of the line is:
$$y - 16 = \frac{16-n}{n-8}(x-n).$$

Since the line passes through the origin, the point $(0,0)$ satisfies the equation of line. So, we have:

$$0 - 16 = \frac{16-n}{n-8}(0-n) \Rightarrow -16(n-8) = (16-n)(-n) \Rightarrow 16 \times 8 = n^2 \Rightarrow n = \pm 8\sqrt{2}.$$

Therefore, one possible value for n is $8\sqrt{2}$, making option A correct.

26) Circumference of the circle is $2 \times \pi \times 7 = 2 \times 3 \times 7 = 42$. Circumference of the square is $10 \times 4 = 40$. Average is:
$$\text{Average} = \frac{42+40}{2} = 41.$$

Therefore, the correct option is D.

27) Let the original price be P. The sale price is $P - 0.25P = 0.75P = \$2,475$. Solving for P:

$$P = \frac{\$2,475}{0.75} = \$3,300.$$

Hence, the original price of the television was $\$3,300$, making option B correct.

28) Perimeter of the semi-circular arch is the sum of the semi-circumference and the diameter:

$$\text{Perimeter} = \pi \times 20 + 2 \times 20 = 62.8 + 40 = 102.8 \; cm.$$

Therefore, the option D is correct.

29) Using the scale, set up a proportion:
$$\frac{0.2}{100} = \frac{x}{5}.$$

Solving for x:
$$x = \frac{0.2 \times 5}{100} = 0.01 \; in.$$

Therefore, 5 feet is $0.01\,in$ in the scale diagram, making option C correct.

11.3 Answers with Explanation

30) First, find X:
$$\frac{4}{9}X = 72 \Rightarrow X = \frac{72 \times 9}{4} = 162.$$

Then find $\frac{3}{4}$ of X:
$$\frac{3}{4} \times 162 = 121.5,$$

therefore, the option E is correct.

31) The distance traveled by the motorcycle is calculated as:

$$\text{Distance} = \text{Speed} \times \text{Time} = 60 \text{ miles/hour} \times 6 \text{ hours} = 360 \text{ miles}.$$

Therefore, the motorcycle will travel 360 miles, making option A correct.

32) The amount spent on the bag is 20% of $40:

$$20\% \times \$40 = 0.20 \times \$40 = \$8.$$

Total spent is the sum of book and bag prices:

$$\$40 + \$8 = \$48.$$

Thus, Lucy spent a total of $48, making option C correct.

33) The smallest factor of 77 (other than 1) is 7. Of the given options, 54 is the only number not divisible by 7, as $54 = 7 \times 7 + 5$ (with a remainder of 5). Therefore, 54 does not have 7 as a factor, making option E correct.

34) The expression simplifies to:
$$\frac{25 + 16 + 36}{9^2} = \frac{77}{81}.$$

Thus, the correct answer is $\frac{77}{81}$, making option A correct.

35) Let the measure of angle D be x. Then, the measure of angle C is $3x$, and they sum to $180°$:

$$x + 3x = 180° \Rightarrow 4x = 180° \Rightarrow x = 45°.$$

Therefore, the measure of angle C is $3 \times 45° = 135°$, making option B correct.

36) Substitute $z = -3$ into the equation:

$$3(-3) + 4 = \frac{w+8}{4}.$$

Simplifying both sides:

$$-9 + 4 = \frac{w+8}{4} \Rightarrow -5 = \frac{w+8}{4}.$$

Solving for w:

$$w + 8 = -20 \Rightarrow w = -20 - 8 = -28.$$

The value of w is -28, making option E correct.

37) Convert heights to inches: Jacob $= 6 \times 12 + 10.5 = 82.5$ inches, Emma $= 5 \times 12 + 4 = 64$ inches. Difference $= 82.5 - 64 = 18.5$ inches, making option A correct.

38) Simplify the expression:

$$\frac{10(y+2) - 15}{15} = \frac{10y + 20 - 15}{15} = \frac{10y + 5}{15} = \frac{2y+1}{3}.$$

Further simplification yields C. $\frac{2}{3}y + \frac{1}{3}$

39) Max completed a total of $15\% + 20\% = 35\%$ of his essay. The fraction left to write is:

$$1 - \frac{35}{100} = \frac{65}{100} = \frac{13}{20}.$$

Which is option A, $\frac{13}{20}$.

40) Using the ratio $3:4$, and knowing there are 12 green markers (representing 4 parts), each part is $\frac{12}{4} = 3$ markers. For red markers (3 parts):

$$3 \times 3 = 9.$$

Thus, there are 9 red markers, making option B correct.

41) Simplify the fraction $-\frac{8}{12}$ to $-\frac{2}{3}$, which as a decimal is $-0.\overline{6}$, option A.

11.3 Answers with Explanation

42) The area A of a circle is given by πr^2, where r is the radius of circle. Set $A = 121\pi$:

$$121\pi = \pi r^2 \Rightarrow r^2 = 121 \Rightarrow r = 11.$$

Therefore, the diameter of the circle is $2 \times 11 = 22$, making option D correct.

43) Sarah's age 10 years ago was $15 - 10 = 5$ years. Therefore, John's age 10 years ago was $4 \times 5 = 20$ years. John's current age is $20 + 10 = 30$ years, which is option C.

44) Consider the prime factorization of 120:

$$120 = 2^3 \times 3 \times 5.$$

Thus, only positive odd factors of 120 are $3, 5, 15$ (the rest factors have 2 as a factor and hence are even). Therefore, there are no positive odd factors of 120 greater than 20 and less than 100, making option A correct.

45) Let the sides be $3x$ and $4x$. The perimeter, $2(3x + 4x) = 56$, gives $14x = 56$. Therefore, $x = 4$, and the sides are $12\ cm$ and $16\ cm$, which is option B.

46) Multiply both sides of the equation by 6, and we have:

$$5(y-3) = 24(\frac{1}{8}y - 1) \Rightarrow 5y - 15 = 3y - 24 \Rightarrow 2y = -9.$$

Solving for y gives $y = -4.5$, making option C correct.

47) The expression $3 - z^2$ reaches its maximum value when z is 0, giving $3 - 0^2 = 3$. Therefore, the option B is correct.

48) Total balls $= 5 + 7 + 9 + 12 = 33$. Probability of picking a red ball $= \frac{5}{33}$, which is option C.

49) Considering the assumptions of the problem, $2x = y$. Thus:

$$2y - 3x = 2(2x) - 3x = 4x - 3x = x = GH.$$

Therefore, the correct answer is option A.

50) To find the percentage of questions Jack answered correctly, first calculate the number of questions he answered correctly:

$$\text{Correct answers} = 60 - 15 = 45.$$

Then, find the percentage of correct answers out of the total questions:

$$\text{Percentage correct} = \left(\frac{45}{60}\right) \times 100\% = 75\%.$$

Therefore, Jack answered 75% of the questions correctly, which corresponds to option C.

Author's Final Note

I hope you enjoyed this book as much as I enjoyed writing it. I have tried to make it as easy to understand as possible. I have also tried to make it fun. I hope I have succeeded. If you have any suggestions for improvement, please let me know. I would love to hear from you.

The accuracy of examples and practice is very important to me. We have done our best. But I also expect that I have made some minor errors. Constant improvement is the name of the game. If you find any errors, please let me know. I will fix them in the next edition.

Your learning journey does not end here. I have written a series of books to help you learn math. Make sure you browse through them. I especially recommend workbooks and practice tests to help you prepare for your exams.

I also enjoy reading your reviews. If you have a moment, please leave a review on Amazon. It will help other students find this book.

If you have any questions or comments, please feel free to contact me at drNazari@effortlessmath.com.

And one last thing: Remember to use online resources for additional help. I recommend using the resources on `https://effortlessmath.com`. There are many great videos on YouTube.

Good luck with your studies!

Dr. Abolfazl Nazari

Author's Final Note

I hope you enjoyed this book as much as I enjoyed writing it. I have tried to make it as easy to understand as possible. I have also tried to make it fun. I hope I have succeeded. If you have any suggestions for improvement, please let me know. I would love to hear from you.

The accuracy of examples and practice is very important to me. We have done our best. But I also expect that I have made some minor errors. Constant improvement is the name of the game. If you find any errors, please let me know. I will fix them in the next edition.

Your learning journey does not end here. I have written a series of books to help you learn math. Make sure you browse through them. I especially recommend workbooks and practice tests to help you prepare for your exams.

I also enjoy reading your reviews. If you have a moment, please leave a review on Amazon. It will help other students find this book.

If you have any questions or comments, please feel free to contact me at drNazari@effortlessmath.com.

And one last thing: Remember to use online resources for additional help. I recommend using the resources on `https://effortlessmath.com`. There are many great videos on YouTube.

Good luck with your studies!

Dr. Abolfazl Nazari